PROBLEM-SOLVING CASES IN
MICROSOFT® ACCESS™ AND EXCEL®

PROBLEM-SOLVING CASES IN MICROSOFT® ACCESS™ AND EXCEL®

Thirteenth Annual Edition

Ellen F. Monk

Joseph A. Brady

Gerard S. Cook

Emilio I. Mendelsohn

CENGAGE
Learning®

Australia • Brazil • Mexico • Singapore • United Kingdom • United States

**Problem-Solving Cases in Microsoft®
Access™ and Excel®, 13th Annual Edition**
Ellen F. Monk, Joseph A. Brady,
Gerard S. Cook, Emilio I. Mendelsohn

Vice President, General Manager, Science,
Math & Quantitative Business: Balraj Kalsi

Product Director: Joe Sabatino

Product Manager: Jason Guyler

Senior Content Developer: Kendra J. Brown

Development Editor: Dan Seiter

Senior Product Assistant: Brad Sullender

Marketing Director: Michele McTighe

Senior Marketing Manager: Eric La Scola

Marketing Coordinator: Will Guiliani

Art and Cover Direction, Production
Management, and Composition: Lumina
Datamatics, Inc.

Media Developer: Chris Valentine

Intellectual Property

    Analyst: Christina Ciaramella

    Project Manager: Betsy Hathaway

Manufacturing Planner: Ron Montgomery

Cover Image(s):

    ©Goodluz/Shutterstock
    ©iStockphoto.com/xavierarnau
    ©merzzie/Shutterstock
    ©Pressmaster/Shutterstock
    ©HlubokiDzianis/Shutterstock

For product information and technology assistance, contact us at
Cengage Learning Customer & Sales Support, 1-800-354-9706
For permission to use material from this text or product,
submit all requests online at **www.cengage.com/permissions**
Further permissions questions can be e-mailed to
permissionrequest@cengage.com

Library of Congress Control Number: 2014957661

Student Edition ISBN: 978-1-305-40872-2

Cengage Learning
20 Channel Center Street
Boston, MA 02210
USA

Cengage Learning is a leading provider of customized learning solutions with office locations around the globe, including Singapore, the United Kingdom, Australia, Mexico, Brazil, and Japan. Locate your local office at: **www.cengage.com/global.**

Cengage Learning products are represented in Canada by Nelson Education, Ltd.

To learn more about Cengage Learning, visit **www.cengage.com.**

Purchase any of our products at your local college store or at our preferred online store: **www.cengagebrain.com.**

Printed in the United States of America
Print Number: 01   Print Year: 2015

To Karen—thank you for the support.

—JAB

To Newton and Olive, who watch over me while I write.

—EFM

To my lovely little Erin, who encourages me to be a better person every day, and to my mother, who did so much to create a better future for me.

—EIM

BRIEF CONTENTS

For more than two decades, we have taught MIS courses at the university level. From the start, we wanted to use good computer-based case studies for the database and decision-support portions of our courses.

At first, we could not find a casebook that met our needs! This surprised us because we thought our requirements were not unreasonable. First, we wanted cases that asked students to think about real-world business situations. Second, we wanted cases that provided students with hands-on experience, using the kind of software that they had learned to use in their computer literacy courses—and that they would later use in business. Third, we wanted cases that would strengthen students' ability to analyze a problem, examine alternative solutions, and implement a solution using software. Undeterred by the lack of casebooks, we wrote our own, and Cengage Learning published it.

This is the thirteenth casebook we have written for Cengage Learning. The cases are all new, and the tutorials have been updated using Microsoft Office 2013.

As with our prior casebooks, we include tutorials that prepare students for the cases, which are challenging but doable. The cases are organized to help students think about the logic of each case's business problem and then about how to use the software to solve the business problem. The cases fit well in an undergraduate MIS course, an MBA information systems course, or a computer science course devoted to business-oriented programming.

BOOK ORGANIZATION

The book is organized into seven parts:

- Database cases using Access
- Decision support cases using the Excel Scenario Manager
- Decision support cases using the Excel Solver
- A decision support case using basic Excel functionality
- Integration cases using Access and Excel
- Advanced Excel skills
- Presentation skills

Part 1 begins with two tutorials that prepare students for the Access case studies. Parts 2 and 3 each begin with a tutorial that prepares students for the Excel case studies. All four tutorials provide students with hands-on practice in using the software's more advanced features—the kind of support that other books about Access and Excel do not provide. Part 4 asks students to use Excel's basic functionality for decision support. Part 5 challenges students to use both Access and Excel to find a solution to a business problem. Part 6 is a tutorial about advanced skills students might need to complete some of the Excel cases. Part 7 is a tutorial that hones students' skills in creating and delivering an oral presentation to business managers. The next sections explore these parts of the book in more depth.

Part 1: Database Cases Using Access

This section begins with two tutorials and then presents five case studies.

Tutorial A: Database Design

This tutorial helps students understand how to set up tables to create a database, without requiring students to learn formal analysis and design methods, such as data normalization.

Tutorial B: Microsoft Access

The second tutorial teaches students the more advanced features of Access queries and reports—features that students will need to know to complete the cases.

Cases 1–5

Five database cases follow Tutorials A and B. The students must use the Access database in each case to create forms, queries, and reports that help management. The first case is an easier "warm-up" case. The next four cases require more effort to design the database and implement the results.

Part 2: Decision Support Cases Using Excel Scenario Manager

This section has one tutorial and two decision support cases that require the use of the Excel Scenario Manager.

Tutorial C: Building a Decision Support System in Excel

This section begins with a tutorial that uses Excel to explain decision support and fundamental concepts of spreadsheet design. The case emphasizes the use of Scenario Manager to organize the output of multiple "what-if" scenarios.

Cases 6–7

Students can complete these two cases with Scenario Manager. In each case, students must use Excel to model two or more solutions to a problem. Students then use the model outputs to identify and document the preferred solution in a memo.

Part 3: Decision Support Cases Using Microsoft Excel Solver

This section has one tutorial and two decision support cases that require the use of Excel Solver.

Tutorial D: Building a Decision Support System Using Microsoft Excel Solver

This section begins with a tutorial for using Excel Solver, a powerful decision support tool for solving optimization problems.

Cases 8–9

Students use the Excel Solver tool in each case to analyze alternatives and identify and document the preferred solution.

Part 4: Decision Support Case Using Basic Excel Functionality

Case 10

The book continues with a case that uses basic Excel functionality. The case does not require Scenario Manager or Solver. Excel is used to test students' analytical skills in a "what-if" analysis.

Part 5: Integration Cases Using Access and Excel

Cases 11 and 12

These cases integrate Access and Excel. The cases show students how to share data between Access and Excel to solve problems.

Part 6: Advanced Skills Using Excel

This part contains one tutorial that focuses on using advanced techniques in Excel.

Tutorial E: Guidance for Excel Cases

Some cases may require the use of Excel techniques that are not discussed in other tutorials or cases in this casebook. For example, techniques for using data tables and pivot tables are explained in Tutorial E rather than in the cases themselves.

Part 7: Presentation Skills
Tutorial F: Giving an Oral Presentation

Each case may includes an optional assignment that lets students practice making a presentation to management to summarize the results of their case analysis. This tutorial gives advice for creating oral presentations. It also includes technical information on charting, a technique that is useful in case analyses or as support for presentations. This tutorial will help students to organize their recommendations, to present their solutions both in words and graphics, and to answer questions from the audience. For larger classes, instructors may want to have students work in teams to create and deliver their presentations, which would model the team approach used by many corporations.

INDIVIDUAL CASE DESIGN

The format of the cases uses the following template:

- Each case begins with a *Preview* and an overview of the tasks.
- The next section, *Preparation*, tells students what they need to do or know to complete the case successfully. Again, the tutorials also prepare students for the cases.
- The third section, *Background*, provides the business context that frames the case. The background of each case models situations that require the kinds of thinking and analysis that students will need in the business world.
- The *Assignment* sections are generally organized to help students develop their analyses.
- The last section, *Deliverables*, lists the finished materials that students must hand in: printouts, a memorandum, a presentation, and files. The list is similar to the deliverables that a business manager might demand.

USING THE CASES

We have successfully used cases like these in our undergraduate MIS courses. We usually begin the semester with Access database instruction. We assign the Access database tutorials and then a case to each student. Then, to teach students how to use the Excel decision support system, we do the same thing: we assign a tutorial and then a case.

Some instructors have asked for access to extra cases, especially in the second semester of a school year. For example, they assigned the integration case in the fall, and they need another one for the spring. To meet this need, we have set up an online "Hall of Fame" that features some of our favorite cases from prior editions. These password-protected cases are available to instructors on the Cengage Learning Web site. Go to *www.cengage.com/login* and search for this textbook by title, author, or ISBN. Note that the cases are in Microsoft Office 2013 format.

TECHNICAL INFORMATION

The cases in this textbook were written using Microsoft Office 2013, and the textbook was tested for quality assurance using the Windows 7 operating system, Microsoft Access 2013, and Microsoft Excel 2013.

Data Files and Solution Files

We have created "starter" data files for the Excel cases, so students need not spend time typing in the spreadsheet skeleton. Cases 11 and 12 also ask students to load Access and Excel starter files. All these files are on the Cengage Learning Web site, which is available both to students and instructors. Instructors should go to *www.cengage.com/login* and search for this textbook by title, author, or ISBN. Students will find the files at *www.cengagebrain.com/login*. You are granted a license to copy the data files to any computer or computer network used by people who have purchased this textbook.

Solutions to the material in the text are available to instructors at *www.cengage.com/login*. Search for this textbook by title, author, or ISBN. The solutions are password protected.

ACKNOWLEDGMENTS

We would like to give many thanks to the team at Cengage Learning, including our Development Editor, Dan Seiter; Senior Content Developer, Kendra Brown; and our Content Project Manager, Divya Divakaran. As always, we acknowledge our students' diligent work.

PART 1

DATABASE CASES USING ACCESS

DATABASE DESIGN

This tutorial has three sections. The first section briefly reviews basic database terminology. The second section teaches database design. The third section features a database design problem for practice.

REVIEW OF TERMINOLOGY

You will begin by reviewing some basic terms that will be used throughout this textbook. In Access, a **database** is a group of related objects that are saved in one file. An Access **object** can be a table, form, query, or report. You can identify an Access database file by its suffix, .accdb.

A **table** consists of data that is arrayed in rows and columns. A **row** of data is called a **record**. A **column** of data is called a **field**. Thus, a record is a set of related fields. The fields in a table should be related to one another in some way. For example, a company might want to keep its employee data together by creating a database table called Employee. That table would contain data fields about employees, such as their names and addresses. It would not have data fields about the company's customers; that data would go in a Customer table.

A field's values have a **data type** that is declared when the table is defined. Thus, when data is entered into the database, the software knows how to interpret each entry. Data types in Access include the following:

- *Text* for words
- *Integer* for whole numbers
- *Double* for numbers that have a decimal value
- *Currency* for numbers that represent dollars and cents
- *Yes/No* for variables that have only two values (such as 1/0, on/off, yes/no, and true/false)
- *Date/Time* for variables that are dates or times

Each database table should have a **primary key** field—a field in which each record has a *unique* value. For example, in an Employee table, a field called Employee Identification Number (EIN) could serve as a primary key. (This assumes that each employee is given a number when hired, and that these numbers are not reused later.) Sometimes, a table does not have a single field whose values are all different. In that case, two or more fields are combined into a **compound primary key**. The combination of the fields' values is unique.

Database tables should be logically related to one another. For example, suppose a company has an Employee table with fields for EIN, Name, Address, and Telephone Number. For payroll purposes, the company has an Hours Worked table with a field that summarizes Labor Hours for individual employees. The relationship between the Employee table and Hours Worked table needs to be established in the database so you can determine the number of hours worked by any employee. To create this relationship, you include the primary key field from the Employee table (EIN) as a field in the Hours Worked table. In the Hours Worked table, the EIN field is then called a **foreign key** because it's from a "foreign" table.

In Access, data can be entered directly into a table or it can be entered into a form, which then inserts the data into a table. A **form** is a database object that is created from an existing table to make the process of entering data more user-friendly.

A **query** is the database equivalent of a question that is posed about data in a table (or tables). For example, suppose a manager wants to know the names of employees who have worked for the company for more than five years. A query could be designed to search the Employee table for the information. The query would be run, and its output would answer the question.

Queries can be designed to search multiple tables at a time. For this to work, the tables must be connected by a **join** operation, which links tables on the values in a field that they have in common. The common field acts as a "hinge" for the joined tables; when the query is run, the query generator treats the joined tables as one large table.

In Access, queries that answer a question are called *select queries* because they select relevant data from the database records. Queries also can be designed to change data in records, add a record to the end of a table, or delete entire records from a table. These queries are called **update**, **append**, and **delete** queries, respectively.

Access has a **report** generator that can be used to format a table's data or a query's output.

DATABASE DESIGN

Designing a database involves determining which tables belong in the database and then creating the fields that belong in each table. This section begins with an introduction to key database design concepts, then discusses design rules you should use when building a database. First, the following key concepts are defined:

- Entities
- Relationships
- Attributes

Database Design Concepts

Computer scientists have highly formalized ways of documenting a database's logic. Learning their notations and mechanics can be time-consuming and difficult. In fact, doing so usually takes a good portion of a systems analysis and design course. This tutorial will teach you database design by emphasizing practical business knowledge; the approach should enable you to design serviceable databases quickly. Your instructor may add more formal techniques.

A database models the logic of an organization's operation, so your first task is to understand the operation. You can talk to managers and workers, make your own observations, and look at business documents such as sales records. Your goal is to identify the business's "entities" (sometimes called *objects*). An **entity** is a thing or event that the database will contain. Every entity has characteristics, called **attributes**, and one or more **relationships** to other entities. Let's take a closer look.

Entities

As previously mentioned, an entity is a tangible thing or an event. The reason for identifying entities is that *an entity eventually becomes a table in the database*. Entities that are things are easy to identify. For example, consider a video store. The database for the video store would probably need to contain the names of DVDs and the names of customers who rent them, so you would have one entity named Video and another named Customer.

In contrast, entities that are events can be more difficult to identify, probably because they are more conceptual. However, events are real, and they are important. In the video store example, one event would be Video Rental and another event would be Hours Worked by employees.

In general, your analysis of an organization's operations is made easier when you realize that organizations usually have physical entities such as these:

- Employees
- Customers
- Inventory (products or services)
- Suppliers

Thus, the database for most organizations would have a table for each of these entities. Your analysis also can be made easier by knowing that organizations engage in transactions internally (within the company) and externally (with the outside world). Such transactions are explained in an introductory accounting course, but most people understand them from events that occur in daily life. Consider the following examples:

- Organizations generate revenue from sales or interest earned. Revenue-generating transactions include event entities called Sales and Interest Earned.
- Organizations incur expenses from paying hourly employees and purchasing materials from suppliers. Hours Worked and Purchases are event entities in the databases of most organizations.

Thus, identifying entities is a matter of observing what happens in an organization. Your powers of observation are aided by knowing what entities exist in the databases of most organizations.

Relationships

As an analyst building a database, you should consider the relationship of each entity to the other entities you have identified. For example, a college database might contain entities for Student, Course, and Section to contain data about each. A relationship between Student and Section could be expressed as "Students enroll in sections."

An analyst also must consider the **cardinality** of any relationship. Cardinality can be one-to-one, one-to-many, or many-to-many:

- In a one-to-one relationship, one instance of the first entity is related to just one instance of the second entity.
- In a one-to-many relationship, one instance of the first entity is related to many instances of the second entity, but each instance of the second entity is related to only one instance of the first.
- In a many-to-many relationship, one instance of the first entity is related to many instances of the second entity, and one instance of the second entity is related to many instances of the first.

For a more concrete understanding of cardinality, consider again the college database with the Student, Course, and Section entities. The university catalog shows that a course such as Accounting 101 can have more than one section: 01, 02, 03, 04, and so on. Thus, you can observe the following relationships:

- The relationship between the entities Course and Section is one-to-many. Each course has many sections, but each section is associated with just one course.
- The relationship between Student and Section is many-to-many. Each student can be in more than one section, because each student can take more than one course. Also, each section has more than one student.

Thinking about relationships and their cardinalities may seem tedious to you. However, as you work through the cases in this text, you will see that this type of analysis can be valuable in designing databases. In the case of many-to-many relationships, you should determine the tables a given database needs; in the case of one-to-many relationships, you should decide which fields the tables need to share.

Attributes

An attribute is a characteristic of an entity. You identify attributes of an entity because *attributes become a table's fields*. If an entity can be thought of as a noun, an attribute can be considered an adjective that describes the noun. Continuing with the college database example, consider the Student entity. Students have names, so Last Name would be an attribute of the Student entity and therefore a field in the Student table. First Name would be another attribute, as well as Address, Phone Number, and other descriptive fields.

Sometimes it can be difficult to tell the difference between an attribute and an entity, but one good way is to ask whether more than one attribute is possible for each entity. If more than one instance is possible, but you do not know the number in advance, you are working with an entity. For example, assume that a student could have a maximum of two addresses—one for home and one for college. You could specify attributes Address 1 and Address 2. Next, consider that you might not know the number of student addresses in advance, meaning that all addresses have to be recorded. In that case, you would not know how many fields to set aside in the Student table for addresses. Therefore, you would need a separate Student Addresses table (entity) that would show any number of addresses for a given student.

Database Design Rules

As described previously, your first task in database design is to understand the logic of the business situation. Once you understand this logic, you are ready to build the database. To create a context for learning about database design, look at a hypothetical business operation and its database needs.

Example: The Talent Agency

Suppose you have been asked to build a database for a talent agency that books musical bands into nightclubs. The agent needs a database to keep track of the agency's transactions and to answer day-to-day questions. For example, a club manager often wants to know which bands are available on a certain date at a certain time, or wants to know the agent's fee for a certain band. The agent may want to see a list of all band members and the instrument each person plays, or a list of all bands that have three members.

Suppose that you have talked to the agent and have observed the agency's business operation. You conclude that your database needs to reflect the following facts:

1. A booking is an event in which a certain band plays in a particular club on a particular date, starting and ending at certain times, and performing for a specific fee. A band can play more than once a day. The Heartbreakers, for example, could play at the East End Cafe in the afternoon and then at the West End Cafe on the same night. For each booking, the club pays the talent agent. The agent keeps a five percent fee and then gives the remainder of the payment to the band.

2. Each band has at least two members and an unlimited maximum number of members. The agent notes a telephone number of just one band member, which is used as the band's contact number. No two bands have the same name or telephone number.

3. Band member names are not unique. For example, two bands could each have a member named Sally Smith.

4. The agent keeps track of just one instrument that each band member plays. For the purpose of this database, "vocals" are considered an instrument.

5. Each band has a desired fee. For example, the Lightmetal band might want $700 per booking, and would expect the agent to try to get at least that amount.

6. Each nightclub has a name, an address, and a contact person. The contact person has a telephone number that the agent uses to call the club. No two clubs have the same name, contact person, or telephone number. Each club has a target fee. The contact person will try to get the agent to accept that fee for a band's appearance.

7. Some clubs feed the band members for free; others do not.

Before continuing with this tutorial, you might try to design the agency's database on your own. Ask yourself: What are the entities? Recall that business databases usually have Customer, Employee, and Inventory entities, as well as an entity for the event that generates revenue transactions. Each entity becomes a table in the database. What are the relationships among the entities? For each entity, what are its attributes? For each table, what is the primary key?

Six Database Design Rules

Assume that you have gathered information about the business situation in the talent agency example. Now you want to identify the tables required for the database and the fields needed in each table. Observe the following six rules:

Rule 1: You do not need a table for the business. The database represents the entire business. Thus, in the example, Agent and Agency are not entities.

Rule 2: Identify the entities in the business description. Look for typical things and events that will become tables in the database. In the talent agency example, you should be able to observe the following entities:

- *Things*: The product (inventory for sale) is Band. The customer is Club.
- *Events*: The revenue-generating transaction is Bookings.

You might ask yourself: Is there an Employee entity? Isn't Instrument an entity? Those issues will be discussed as the rules are explained.

Rule 3: Look for relationships among the entities. Look for one-to-many relationships between entities. The relationship between those entities must be established in the tables, using a foreign key. For details, see the following discussion in Rule 4 about the relationship between Band and Band Member.

Look for many-to-many relationships between entities. Each of these relationships requires a third entity that associates the two entities in the relationship. Recall the many-to-many relationship from the college database scenario that involved Student and Section entities. To display the enrollment of specific students in specific sections, a third table would be required. The mechanics of creating such a table are described in Rule 4 during the discussion of the relationship between Band and Club.

Rule 4: Look for attributes of each entity and designate a primary key. As previously mentioned, you should think of the entities in your database as nouns. You should then create a list of adjectives that describe those nouns. These adjectives are the attributes that will become the table's fields. After you have identified fields for each table, you should check to see whether a field has unique values. If such a field exists, designate it as the primary key field; otherwise, designate a compound primary key.

In the talent agency example, the attributes, or fields, of the Band entity are Band Name, Band Phone Number, and Desired Fee, as shown in Figure A-1. Assume that no two bands have the same name, so the primary key field can be Band Name. The data type of each field is shown.

BAND	
Field Name	Data Type
Band Name (primary key)	Text
Band Phone Number	Text
Desired Fee	Currency

Source: © 2015 Cengage Learning®

FIGURE A-1 The Band table and its fields

Two Band records are shown in Figure A-2.

Band Name (primary key)	Band Phone Number	Desired Fee
Heartbreakers	981 831 1765	$800
Lightmetal	981 831 2000	$700

Source: © 2015 Cengage Learning®

FIGURE A-2 Records in the Band table

If two bands might have the same name, Band Name would not be a good primary key, so a different unique identifier would be needed. Such situations are common. Most businesses have many types of inventory, and duplicate names are possible. The typical solution is to assign a number to each product to use as the primary key field. A college could have more than one faculty member with the same name, so each faculty member would be assigned an employee identification number. Similarly, banks assign a personal identification number (PIN) for each depositor. Each automobile produced by a car manufacturer gets a unique Vehicle Identification Number (VIN). Most businesses assign a number to each sale, called an invoice number. (The next time you go to a grocery store, note the number on your receipt. It will be different from the number on the next customer's receipt.)

At this point, you might be wondering why Band Member would not be an attribute of Band. The answer is that, although you must record each band member, you do not know in advance how many members are in each band. Therefore, you do not know how many fields to allocate to the Band table for members. (Another way to think about band members is that they are the agency's employees, in effect. Databases for organizations usually have an Employee entity.) You should create a Band Member table with the attributes Member ID Number, Member Name, Band Name, Instrument, and Phone. A Member ID Number field is needed because member names may not be unique. The table and its fields are shown in Figure A-3.

BAND MEMBER	
Field Name	Data Type
Member ID Number (primary key)	Text
Member Name	Text
Band Name (foreign key)	Text
Instrument	Text
Phone	Text

Source: © 2015 Cengage Learning®

FIGURE A-3 The Band Member table and its fields

Note in Figure A-3 that the phone number is classified as a Text data type because the field values will not be used in an arithmetic computation. The benefit is that Text data type values take up fewer bytes than Numerical or Currency data type values; therefore, the file uses less storage space. You should also use the Text data type for number values such as zip codes.

Five records in the Band Member table are shown in Figure A-4.

Member ID Number (primary key)	Member Name	Band Name	Instrument	Phone
0001	Pete Goff	Heartbreakers	Guitar	981 444 1111
0002	Joe Goff	Heartbreakers	Vocals	981 444 1234
0003	Sue Smith	Heartbreakers	Keyboard	981 555 1199
0004	Joe Jackson	Lightmetal	Sax	981 888 1654
0005	Sue Hoopes	Lightmetal	Piano	981 888 1765

Source: © 2015 Cengage Learning®

FIGURE A-4   Records in the Band Member table

You can include Instrument as a field in the Band Member table because the agent records only one instrument for each band member. Thus, you can use the instrument as a way to describe a band member, much like the phone number is part of the description. Phone could not be the primary key because two members might share a telephone and because members might change their numbers, making database administration more difficult.

You might ask why Band Name is included in the Band Member table. The common-sense reason is that you did not include the Member Name in the Band table. You must relate bands and members somewhere, and the Band Member table is the place to do it.

To think about this relationship in another way, consider the cardinality of the relationship between Band and Band Member. It is a one-to-many relationship: one band has many members, but each member in the database plays in just one band. You establish such a relationship in the database by using the primary key field of one table as a foreign key in the other table. In Band Member, the foreign key Band Name is used to establish the relationship between the member and his or her band.

The attributes of the Club entity are Club Name, Address, Contact Name, Club Phone Number, Preferred Fee, and Feed Band?. The Club table can define the Club entity, as shown in Figure A-5.

CLUB	
Field Name	Data Type
Club Name (primary key)	Text
Address	Text
Contact Name	Text
Club Phone Number	Text
Preferred Fee	Currency
Feed Band?	Yes/No

Source: © 2015 Cengage Learning®

FIGURE A-5   The Club table and its fields

Two records in the Club table are shown in Figure A-6.

Club Name (primary key)	Address	Contact Name	Club Phone Number	Preferred Fee	Feed Band?
East End	1 Duce St.	Al Pots	981 444 8877	$600	Yes
West End	99 Duce St.	Val Dots	981 555 0011	$650	No

Source: © 2015 Cengage Learning®

FIGURE A-6   Records in the Club table

You might wonder why Bands Booked into Club (or a similar name) is not an attribute of the Club table. There are two reasons. First, you do not know in advance how many bookings a club will have, so the value cannot be an attribute. Second, Bookings is the agency's revenue-generating transaction, an event entity, and you need a table for that business transaction. Consider the booking transaction next.

You know that the talent agent books a certain band into a certain club for a specific fee on a certain date, starting and ending at a specific time. From that information, you can see that the attributes of the Bookings entity are Band Name, Club Name, Date, Start Time, End Time, and Fee. The Bookings table and its fields are shown in Figure A-7.

BOOKINGS	
Field Name	**Data Type**
Band Name (foreign key)	Text
Club Name (foreign key)	Text
Date	Date/Time
Start Time	Date/Time
End Time	Date/Time
Fee	Currency

Source: © 2015 Cengage Learning®

FIGURE A-7   The Bookings table and its fields—and no designation of a primary key

Some records in the Bookings table are shown in Figure A-8.

Band Name	Club Name	Date	Start Time	End Time	Fee
Heartbreakers	East End	11/21/14	21:30	23:30	$800
Heartbreakers	East End	11/22/14	21:00	23:30	$750
Heartbreakers	West End	11/28/14	19:00	21:00	$500
Lightmetal	East End	11/21/14	18:00	20:00	$700
Lightmetal	West End	11/22/14	19:00	21:00	$750

Source: © 2015 Cengage Learning®

FIGURE A-8   Records in the Bookings table

Note that no single field is guaranteed to have unique values, because each band is likely to be booked many times and each club might be used many times. Furthermore, each date and time can appear more than once. Thus, no one field can be the primary key.

If a table does not have a single primary key field, you can make a compound primary key whose field values will be unique when taken together. Because a band can be in only one place at a time, one possible solution is to create a compound key from the Band Name, Date, and Start Time fields. An alternative solution is to create a compound primary key from the Club Name, Date, and Start Time fields.

If you don't want a compound key, you could create a field called Booking Number. Each booking would then have its own unique number, similar to an invoice number.

You can also think about this event entity in a different way. Over time, a band plays in many clubs, and each club hires many bands. Thus, Band and Club have a many-to-many relationship, which signals the need for a table between the two entities. A Bookings table would associate the Band and Club tables. You implement an associative table by including the primary keys from the two tables that are associated. In this case, the primary keys from the Band and Club tables are included as foreign keys in the Bookings table.

Rule 5: *Avoid data redundancy.* You should not include extra (redundant) fields in a table. Redundant fields take up extra disk space and lead to data entry errors because the same value must be entered in multiple tables, increasing the chance of a keystroke error. In large databases, keeping track of multiple instances of the same data is nearly impossible, so contradictory data entries become a problem.

Consider this example: Why wouldn't Club Phone Number be included in the Bookings table as a field? After all, the agent might have to call about a last-minute booking change and could quickly look up the number in the Bookings table. Assume that the Bookings table includes Booking Number as the primary key and Club Phone Number as a field. Figure A-9 shows the Bookings table with the additional field.

BOOKINGS	
Field Name	**Data Type**
Booking Number (primary key)	Text
Band Name (foreign key)	Text
Club Name (foreign key)	Text
Club Phone Number	Text
Date	Date/Time
Start Time	Date/Time
End Time	Date/Time
Fee	Currency

Source: © 2015 Cengage Learning®

FIGURE A-9   The Bookings table with an unnecessary field—Club Phone Number

The fields Date, Start Time, End Time, and Fee logically depend on the Booking Number primary key—they help define the booking. Band Name and Club Name are foreign keys and are needed to establish the relationship between the Band, Club, and Bookings tables. But what about Club Phone Number? It is not defined by the Booking Number. It is defined by Club Name—*in other words, it is a function of the club, not of the booking.* Thus, the Club Phone Number field does not belong in the Bookings table. It is already in the Club table.

Perhaps you can see the practical data-entry problem of including Club Phone Number in Bookings. Suppose a club changed its contact phone number. The agent could easily change the number one time, in the Club table. However, the agent would need to remember which other tables contained the field and change the values there too. In a small database, this task might not be difficult, but in larger databases, having redundant fields in many tables makes such maintenance difficult, which means that redundant data is often incorrect.

You might object by saying, "What about all of those foreign keys? Aren't they redundant?" In a sense, they are. But they are needed to establish the one-to-many relationship between one entity and another, as discussed previously.

Rule 6: *Do not include a field if it can be calculated from other fields.* A calculated field is made using the query generator. Thus, the agent's fee is not included in the Bookings table because it can be calculated by query (here, five percent multiplied by the booking fee).

PRACTICE DATABASE DESIGN PROBLEM

Imagine that your town library wants to keep track of its business in a database, and that you have been called in to build the database. You talk to the town librarian, review the old paper-based records, and watch people use the library for a few days. You learn the following about the library:

1. Any resident of the town can get a library card simply by asking for one. The library considers each cardholder a member of the library.

2. The librarian wants to be able to contact members by telephone and by mail. She calls members when books are overdue or when requested materials become available. She likes to mail a thank-you note to each patron on his or her anniversary of becoming a member of the library. Without a database, contacting members efficiently can be difficult; for example, multiple members can have the same name. Also, a parent and a child might have the same first and last name, live at the same address, and share a phone.

3. The librarian tries to keep track of each member's reading interests. When new books come in, the librarian alerts members whose interests match those books. For example, long-time member Sue Doaks is interested in reading Western novels, growing orchids, and baking bread. There must be some way to match her interests with available books. One complication is that, although the librarian wants to track all of a member's reading interests, she wants to classify each book as being in just one category of interest. For example, the classic gardening book *Orchids of France* would be classified as a book about orchids or a book about France, but not both.

4. The library stocks thousands of books. Each book has a title and any number of authors. Also, more than one book in the library might have the same title. Similarly, multiple authors might have the same name.

5. A writer could be the author of more than one book.

6. A book will be checked out repeatedly as time goes on. For example, *Orchids of France* could be checked out by one member in March, by another member in July, and by another member in September.

7. The library must be able to identify whether a book is checked out.

8. A member can check out any number of books in one visit. Also, a member might visit the library more than once a day to check out books.

9. All books that are checked out are due back in two weeks, with no exceptions. The librarian would like to have an automated way of generating an overdue book list each day so she can telephone offending members.

10. The library has a number of employees. Each employee has a job title. The librarian is paid a salary, but other employees are paid by the hour. Employees clock in and out each day. Assume that all employees work only one shift per day and that all are paid weekly. Pay is deposited directly into an employee's checking account—no checks are hand-delivered. The database needs to include the librarian and all other employees.

Design the library's database, following the rules set forth in this tutorial. Your instructor will specify the format of your work. Here are a few hints in the form of questions:

- A book can have more than one author. An author can write more than one book. How would you describe the relationship between books and authors?

- The library lends books for free, of course. If you were to think of checking out a book as a sales transaction for zero revenue, how would you handle the library's revenue-generating event?

- A member can borrow any number of books at one checkout. A book can be checked out more than once. How would you describe the relationship between checkouts and books?

MICROSOFT ACCESS

Microsoft Access is a relational database package that runs on the Microsoft Windows operating system. There are many different versions of Access; this tutorial was prepared using Access 2013.

Before using this tutorial, you should know the fundamentals of Access and know how to use Windows. This tutorial explains advanced Access skills you will need to complete database case studies. The tutorial concludes with a discussion of common Access problems and how to solve them.

To prevent losing your work, always observe proper file-saving and closing procedures. To exit Access, click the File tab and select Close, then click the Close button in the upper-right corner. Always end your work with these steps. If you remove your USB key or other portable storage device when database forms and tables are shown on the screen, you will lose your work.

To begin this tutorial, you will create a new database called Employee.

AT THE KEYBOARD

Open Access. Click the Blank desktop database icon from the templates list. Name the database Employee. Click the file folder next to the filename to browse for the folder where you want to save the file. Click the new folder, click Open, and then click OK. Otherwise, your file will be saved automatically in the Documents folder. Click the Create button.

A portion of your opening screen should resemble the screen shown in Figure B-1.

Source: Microsoft product screenshots used with permission from Microsoft Corporation

FIGURE B-1   Entering data in Datasheet view

When you create a table, Access opens it in Datasheet view by default. Because you will use Design view to build your tables, close the new table by clicking the *X* in the upper-right corner of the table window that corresponds to Close 'Table1.' You are now on the Home tab in the Database window of Access, as shown in Figure B-2. From this screen, you can create or change objects.

Source: Microsoft product screenshots used with permission from Microsoft Corporation

FIGURE B-2   The Database window Home tab in Access

CREATING TABLES

Your database will contain data about employees, their wage rates, and the hours they worked.

Defining Tables

In the Database window, build three new tables using the following instructions.

AT THE KEYBOARD

Defining the Employee Table

This table contains permanent data about employees. To create the table, click the Create tab and then click Table Design in the Tables group. The table's fields are Last Name, First Name, Employee ID, Street Address, City, State, Zip, Date Hired, and US Citizen. The Employee ID field is the primary key field. Change the lengths of Short Text fields from the default 255 spaces to more appropriate lengths; for example, the Last Name field might be 30 spaces, and the Zip field might be 10 spaces. Your completed definition should resemble the one shown in Figure B-3.

Field Name	Data Type	Description (Optional)
Last Name	Short Text	
First Name	Short Text	
⚲ Employee ID	Short Text	
Street Address	Short Text	
City	Short Text	
State	Short Text	
Zip	Short Text	
Date Hired	Date/Time	
US Citizen	Yes/No	

Source: Microsoft product screenshots used with permission from Microsoft Corporation

FIGURE B-3   Fields in the Employee table

When you finish, click the File tab, select Save As, select Save Object As, click the Save As button, and then enter a name for the table. In this example, the table is named Employee. (It is a coincidence that the Employee table has the same name as its database file.) After entering the name, click OK in the Save As window. Close the table by clicking the Close button (X) that corresponds to the Employee table.

Defining the Wage Data Table

This table contains permanent data about employees and their wage rates. The table's fields are Employee ID, Wage Rate, and Salaried. The Employee ID field is the primary key field. Use the data types shown in Figure B-4. Your definition should resemble the one shown in Figure B-4.

Field Name	Data Type	Description (Optional)
⚲ Employee ID	Short Text	
Wage Rate	Currency	
Salaried	Yes/No	

Source: Microsoft product screenshots used with permission from Microsoft Corporation

FIGURE B-4   Fields in the Wage Data table

Click the File tab and then select Save As, select Save Object As, and click the Save As button to save the table definition. Name the table Wage Data.

Defining the Hours Worked Table

The purpose of this table is to record the number of hours that employees work each week during the year. The table's three fields are Employee ID (which has a Short Text data type), Week # (number–long integer), and Hours (number–double). The Employee ID and Week # are the compound keys.

In the following example, the employee with ID number 08965 worked 40 hours in Week 1 of the year and 52 hours in Week 2.

Employee ID	Week #	Hours
08965	1	40
08965	2	52

Note that no single field can be the primary key field because 08965 is an entry for each week. In other words, if this employee works each week of the year, 52 records will have the same Employee ID value at the end of the year. Thus, Employee ID values will not distinguish records. No other single field can distinguish these records either, because other employees will have worked during the same week number and some employees will have worked the same number of hours. For example, 40 hours—which corresponds to a full-time workweek—would be a common entry for many weeks.

All of this presents a problem because a table must have a primary key field in Access. The solution is to use a compound primary key; that is, use values from more than one field to create a combined field that will distinguish records. The best compound key to use for the current example consists of the Employee ID field and the Week # field, because as each person works each week, the week number changes. In other words, there is only *one* combination of Employee ID 08965 and Week # 1. Because those values *can occur in only one record*, the combination distinguishes that record from all others.

The first step of setting a compound key is to highlight the fields in the key. Those fields must appear one after the other in the table definition screen. (Plan ahead for that format.) As an alternative, you can highlight one field, hold down the Control key, and highlight the next field.

AT THE KEYBOARD

In the Hours Worked table, click the first field's left prefix area (known as the row selector), hold down the mouse button, and drag down to highlight the names of all fields in the compound primary key. Your screen should resemble the one shown in Figure B-5.

Table1		
Field Name	Data Type	Description (Optional)
Employee ID	Short Text	
Week #	Number	
Hours	Number	

Source: Microsoft product screenshots used with permission from Microsoft Corporation

FIGURE B-5   Selecting fields for the compound primary key for the Hours Worked table

Now click the Key icon. Your screen should resemble the one shown in Figure B-6.

Table1		
Field Name	Data Type	Description (Optional)
Employee ID	Short Text	
Week #	Number	
Hours	Number	

Source: Microsoft product screenshots used with permission from Microsoft Corporation

FIGURE B-6   The compound primary key for the Hours Worked table

You have created the compound primary key and finished defining the table. Click the File tab and then select Save As, select Save Object As, and click the Save As button to save the table as Hours Worked.

Adding Records to a Table

At this point, you have set up the skeletons of three tables. The tables have no data records yet. If you printed the tables now, you would only see column headings (the field names). The most direct way to enter data into a table is to double-click the table's name in the navigation pane at the left side of the screen and then type the data directly into the cells.

NOTE

To display and open the database objects, Access 2013 uses a navigation pane, which is on the left side of the Access window.

AT THE KEYBOARD

On the Home tab of the Database window, double-click the Employee table. Your data entry screen should resemble the one shown in Figure B-7.

Employee								
Last Name ▾	First Name ▾	Employee ID ▾	Street Addre ▾	City ▾	State ▾	Zip ▾	Date Hired ▾	US Citizen ▾ C
*								☐

Source: Microsoft product screenshots used with permission from Microsoft Corporation

FIGURE B-7   The data entry screen for the Employee table

The Employee table has many fields, some of which may be off the screen to the right. Scroll to see obscured fields. (Scrolling happens automatically as you enter data.) Figure B-7 shows all of the fields on the screen.

Enter your data one field value at a time. Note that the first row is empty when you begin. Each time you finish entering a value, press Enter to move the cursor to the next cell. After you enter data in the last cell in a row, the cursor moves to the first cell of the next row *and* Access automatically saves the record. Thus, you do not need to click the File tab and then select Save after entering data into a table.

When entering data in your table, you should enter dates in the following format: 6/15/10. Access automatically expands the entry to the proper format in output.

Also note that Yes/No variables are clicked (checked) for Yes; otherwise, the box is left blank for No. You can change the box from Yes to No by clicking it.

Enter the data shown in Figure B-8 into the Employee table. If you make errors in data entry, click the cell, backspace over the error, and type the correction.

Employee									
Last Name ▾	First Name ▾	Employee ID ▾	Street Address ▾	City ▾	State ▾	Zip ▾	Date Hired ▾	US Citizen ▾	Click to Add
Howard	Jane	11411	28 Sally Dr	Glasgow	DE	19702	6/1/2015	☑	
Johnson	John	12345	30 Elm St	Newark	DE	19711	6/1/1996	☑	
Smith	Albert	14890	44 Duce St	Odessa	DE	19722	7/15/1987	☑	
Jones	Sue	22282	18 Spruce St	Newark	DE	19716	7/15/2004	☐	
Ruth	Billy	71460	1 Tater Dr	Baltimore	MD	20111	8/15/1999	☐	
Add	Your	Data	Here	Elkton	MD	21921		☑	
*								☐	

Source: Microsoft product screenshots used with permission from Microsoft Corporation

FIGURE B-8   Data for the Employee table

Note that the sixth record is *your* data record. Assume that you live in Elkton, Maryland, were hired on today's date (enter the date), and are a U.S. citizen. Make up a fictitious Employee ID number. For purposes of this tutorial, the sixth record has been created using the name of one of this text's authors and the employee ID 09911.

After adding records to the Employee table, open the Wage Data table and enter the data shown in Figure B-9.

Wage Data		
Employee ID ▾	Wage Rate ▾	Salaried ▾
11411	$10.00	☐
12345	$0.00	☑
14890	$12.00	☐
22282	$0.00	☑
71460	$0.00	☑
Your Employee ID	$8.00	☐
*	$0.00	☐

Source: Microsoft product screenshots used with permission from Microsoft Corporation

FIGURE B-9   Data for the Wage Data table

In this table, you are again asked to create a new entry. For this record, enter your own employee ID. Also assume that you earn $8 an hour and are not salaried. Note that when an employee's Salaried box is not checked (in other words, Salaried = No), the implication is that the employee is paid by the hour. Because salaried employees are not paid by the hour, their hourly rate is 0.00.

When you finish creating the Wage Data table, open the Hours Worked table and enter the data shown in Figure B-10.

Hours Worked		
Employee ID	Week #	Hours
11411	1	40
11411	2	50
12345	1	40
12345	2	40
14890	1	38
14890	2	40
22282	1	40
22282	2	40
71460	1	40
71460	2	40
Your Employee ID	1	60
Your Employee ID	2	55
*	0	0

Source: Microsoft product screenshots used with permission from Microsoft Corporation

FIGURE B-10   Data for the Hours Worked table

Notice that salaried employees are always given 40 hours. Nonsalaried employees (including you) might work any number of hours. For your record, enter your fictitious employee ID, 60 hours worked for Week 1, and 55 hours worked for Week 2.

CREATING QUERIES

Because you know how to create basic queries, this section explains the advanced queries you will create in the cases in this book.

Using Calculated Fields in Queries

A **calculated field** is an output field made up of *other* field values. A calculated field can be a field in a table; here it is created in the query generator. The calculated field here does not become part of the table—it is just part of the query output. The best way to understand this process is to work through an example.

AT THE KEYBOARD

Suppose you want to see the employee IDs and wage rates of hourly workers, and the new wage rates if all employees were given a 10 percent raise. To view that information, show the employee ID, the current wage rate, and the higher rate, which should be titled New Rate in the output. Figure B-11 shows how to set up the query.

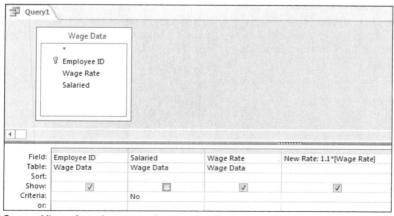

Field:	Employee ID	Salaried	Wage Rate	New Rate: 1.1*[Wage Rate]
Table:	Wage Data	Wage Data	Wage Data	
Sort:				
Show:	✓	☐	✓	✓
Criteria:		No		
or:				

Source: Microsoft product screenshots used with permission from Microsoft Corporation

FIGURE B-11   Query setup for the calculated field

To set up this query, you need to select hourly workers by using the Salaried field with Criteria = No. Note in Figure B-11 that the Show box for the field is not checked, so the Salaried field values will not appear in the query output.

Note the expression for the calculated field, which you can see in the far-right field cell:

New Rate: 1.1 * [Wage Rate]

The term *New Rate:* merely specifies the desired output heading. (Don't forget the colon.) The rest of the expression, 1.1 * [Wage Rate], multiplies the old wage rate by 110 percent, which results in the 10 percent raise.

In the expression, the field name Wage Rate must be enclosed in square brackets. Remember this rule: *Any time an Access expression refers to a field name, the field name must be enclosed in square brackets*.

If you run this query, your output should resemble that in Figure B-12.

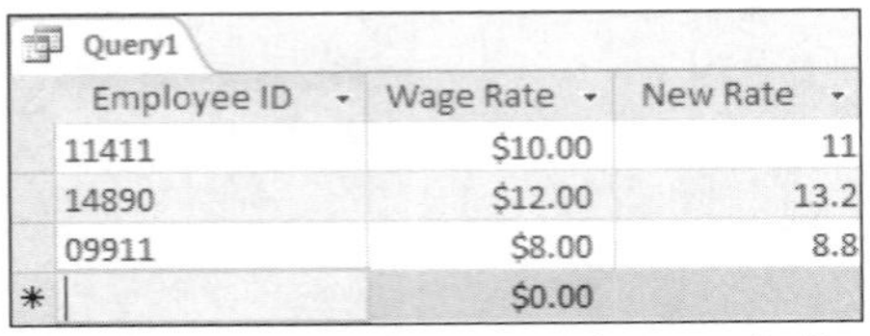

Query1

Employee ID	Wage Rate	New Rate
11411	$10.00	11
14890	$12.00	13.2
09911	$8.00	8.8
*	$0.00	

Source: Microsoft product screenshots used with permission from Microsoft Corporation

FIGURE B-12 Output for a query with calculated field

Notice that the calculated field output is not shown in Currency format, but as a Double—a number with digits after the decimal point. To convert the output to Currency format, select the output column by clicking the line above the calculated field expression. The column darkens to indicate its selection. Your data entry screen should resemble the one shown in Figure B-13.

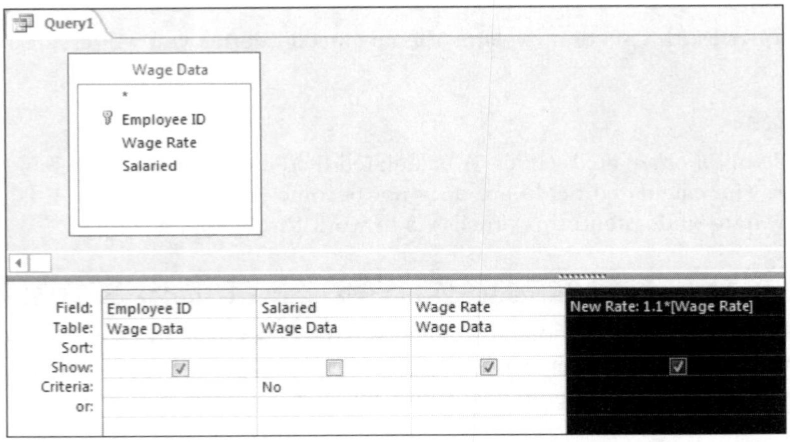

Source: Microsoft product screenshots used with permission from Microsoft Corporation

FIGURE B-13 Activating a calculated field in query design

Then, on the Design tab, click Property Sheet in the Show/Hide group. The Field Properties sheet appears, as shown on the right in Figure B-14.

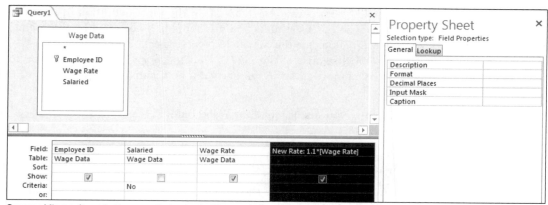

Source: Microsoft product screenshots used with permission from Microsoft Corporation

FIGURE B-14   Field properties of a calculated field

Click Format and choose Currency, as shown in Figure B-15. Then click the X in the upper-right corner of the window to close it.

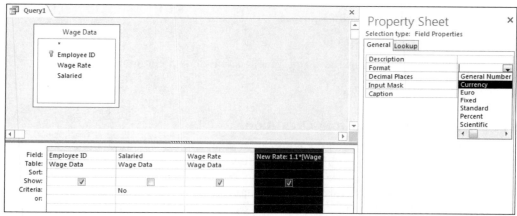

Source: Microsoft product screenshots used with permission from Microsoft Corporation

FIGURE B-15   Currency format of a calculated field

When you run the query, the output should resemble that in Figure B-16.

Employee ID ▾	Wage Rate ▾	New Rate ▾
11411	$10.00	$11.00
14890	$12.00	$13.20
09911	$8.00	$8.80
*		$0.00

Source: Microsoft product screenshots used with permission from Microsoft Corporation

FIGURE B-16   Query output with formatted calculated field

Next, you examine how to avoid errors when making calculated fields.

Avoiding Errors when Making Calculated Fields

Follow these guidelines to avoid making errors in calculated fields:

- Do not enter the expression in the *Criteria* cell as if the field definition were a filter. You are making a field, so enter the expression in the *Field* cell.

- Spell, capitalize, and space a field's name *exactly* as you did in the table definition. If the table definition differs from what you type, Access thinks you are defining a new field by that name. Access then prompts you to enter values for the new field, which it calls a Parameter Query field. This problem is easy to debug because of the tag *Parameter Query*. If Access asks you to enter values for a parameter, you almost certainly misspelled a field name in an expression in a calculated field or criterion.

  For example, here are some errors you might make for Wage Rate:

  > Misspelling: (Wag Rate)
  > Case change: (wage Rate / WAGE RATE)
  > Spacing change: (WageRate / Wage Rate)

- Do not use parentheses or curly braces instead of the square brackets. Also, do not put parentheses inside square brackets. You *can*, however, use parentheses outside the square brackets in the normal algebraic manner.

  For example, suppose that you want to multiply Hours by Wage Rate to get a field called Wages Owed. This is the correct expression:

  > Wages Owed: [Wage Rate] * [Hours]

  The following expression also would be correct:

  > Wages Owed: ([Wage Rate] * [Hours])

  But it would *not* be correct to omit the inside brackets, which is a common error:

  > Wages Owed: [Wage Rate * Hours]

"Relating" Two or More Tables by the Join Operation

Often, the data you need for a query is in more than one table. To complete the query, you must **join** the tables by linking the common fields. One rule of thumb is that joins are made on fields that have common *values*, and those fields often can be key fields. The names of the join fields are irrelevant; also, the names of the tables or fields to be joined may be the same, but it is not required for an effective join.

Make a join by bringing in (adding) the tables needed. Next, decide which fields you will join. Then click one field name and hold down the left mouse button while you drag the cursor over to the other field's name in its window. Release the button. Access inserts a line to signify the join. (If a relationship between two tables has been formed elsewhere, Access inserts the line automatically, and you do not have to perform the click-and-drag operation. Access often inserts join lines without the user forming relationships.)

You can join more than two tables. The common fields *need not* be the same in all tables; that is, you can daisy-chain them together.

A common join error is to add a table to the query and then fail to link it to another table. In that case, you will have a table floating in the top part of the QBE (query by example) screen. When you run the query, your output will show the same records over and over. The error is unmistakable because there is *so much* redundant output. The two rules are to add only the tables you need and to link all tables.

Next, you will work through an example of a query that needs a join.

AT THE KEYBOARD

Suppose you want to see the last names, employee IDs, wage rates, salary status, and citizenship only for U.S. citizens and hourly workers. Because the data is spread across two tables, Employee and Wage Data, you should add both tables and pull down the five fields you need. Then you should add the Criteria expressions. Set up your work to resemble that in Figure B-17. Make sure the tables are joined on the common field, Employee ID.

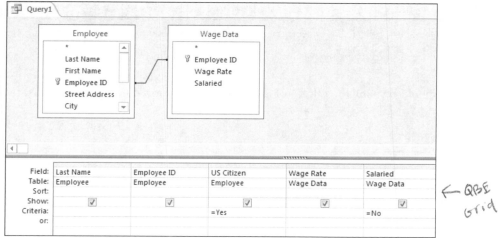

Source: Microsoft product screenshots used with permission from Microsoft Corporation

FIGURE B-17 A query based on two joined tables

You should quickly review the criteria you will need to set up this join: If you want data for employees who are U.S. citizens *and* who are hourly workers, the Criteria expressions go in the *same* Criteria row. If you want data for employees who are U.S. citizens *or* who are hourly workers, one of the expressions goes in the second Criteria row (the one with the or: notation).

Now run the query. The output should resemble that in Figure B-18, with the exception of the name "Brady."

Last Name	Employee ID	US Citizen	Wage Rate	Salaried
Howard	11411	✓	$10.00	☐
Smith	14890	✓	$12.00	☐
Brady	09911	✓	$8.00	☐

Source: Microsoft product screenshots used with permission from Microsoft Corporation

FIGURE B-18 Output of a query based on two joined tables

You do not need to print or save the query output, so return to Design view and close the query. Another practice query follows.

AT THE KEYBOARD

Suppose you want to see the wages owed to hourly employees for Week 2. You should show the last name, the employee ID, the salaried status, the week #, and the wages owed. Wages will have to be a calculated field ([Wage Rate] * [Hours]). The criteria are No for Salaried and 2 for the Week #. (This means that another "And" query is required.) Your query should be set up like the one in Figure B-19.

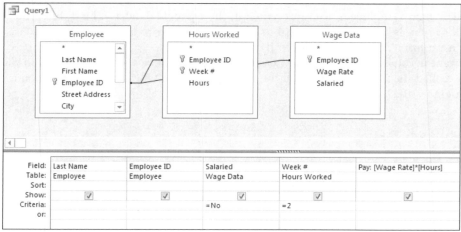

Source: Microsoft product screenshots used with permission from Microsoft Corporation

FIGURE B-19 Query setup for wages owed to hourly employees for Week 2

NOTE

In the query in Figure B-19, the calculated field column was widened so you could see the whole expression. To widen a column, click the column boundary line and drag to the right.

Run the query. The output should be similar to that in Figure B-20, if you formatted your calculated field to Currency.

Last Name	Employee ID	Salaried	Week #	Pay
Howard	11411	☐	2	$500.00
Smith	14890	☐	2	$480.00
Brady	09911	☐	2	$440.00
*		▣		

Source: Microsoft product screenshots used with permission from Microsoft Corporation

FIGURE B-20   Query output for wages owed to hourly employees for Week 2

Notice that it was not necessary to pull down the Wage Rate and Hours fields to make the query work. You do not need to save or print the query output, so return to Design view and close the query.

Summarizing Data from Multiple Records (Totals Queries)

You may want data that summarizes values from a field for several records (or possibly all records) in a table. For example, you might want to know the average hours that all employees worked in a week or the total (sum) of all of the hours worked. Furthermore, you might want data grouped or stratified in some way. For example, you might want to know the average hours worked, grouped by all U.S. citizens versus all non-U.S. citizens. Access calls such a query a **Totals query**. These queries include the following operations:

Sum	The total of a given field's values
Count	A count of the number of instances in a field—that is, the number of records. In the current example, you would count the number of employee IDs to get the number of employees.
Average	The average of a given field's values
Min	The minimum of a given field's values
Var	The variance of a given field's values
StDev	The standard deviation of a given field's values
Where	The field has criteria for the query output

AT THE KEYBOARD

Suppose you want to know how many employees are represented in the example database. First, bring the Employee table into the QBE screen. Because you will need to count the number of employee IDs, which is a Totals query operation, you must bring down the Employee ID field.

To tell Access that you want a Totals query, click the Design tab and then click the Totals button in the Show/Hide group. A new row called the Total row opens in the lower part of the QBE screen. At this point, the screen resembles that in Figure B-21.

Count operation

Totals query → Σ (sum)

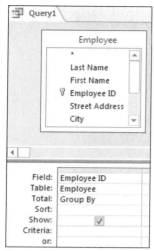

Source: Microsoft product screenshots used with permission from Microsoft Corporation

FIGURE B-21   Totals query setup

Note that the Total cell contains the words *Group By*. Until you specify a statistical operation, Access assumes that a field will be used for grouping (stratifying) data.

To count the number of employee IDs, click next to Group By to display an arrow. Click the arrow to reveal a drop-down menu, as shown in Figure B-22.

Source: Microsoft product screenshots used with permission from Microsoft Corporation

FIGURE B-22   Choices for statistical operation in a Totals query

Select the Count operator. (You might need to scroll down the menu to see the operator you want.) Your screen should resemble the one shown in Figure B-23.

FIGURE B-23   Count in a Totals query

Run the query. Your output should resemble that in Figure B-24.

FIGURE B-24   Output of Count in a Totals query

Notice that Access created a pseudo-heading, "CountOfEmployee ID," by splicing together the statistical operation (Count), the word Of, and the name of the field (Employee ID). If you wanted a phrase such as "Count of Employees" as a heading, you would go to Design view and change the query to resemble the one shown in Figure B-25.

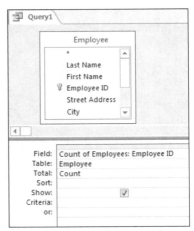

FIGURE B-25   Heading change in a Totals query

When you run the query, the output should resemble that in Figure B-26.

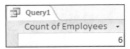

FIGURE B-26   Output of heading change in a Totals query

You do not need to print or save the query output, so return to Design view and close the query.

AT THE KEYBOARD

As another example of a Totals query, suppose you want to know the average wage rate of employees, grouped by whether the employees are salaried. Figure B-27 shows how to set up your query.

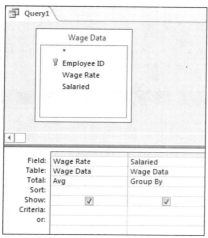

Source: Microsoft product screenshots used with permission from Microsoft Corporation

FIGURE B-27   Query setup for average wage rate of employees

When you run the query, your output should resemble that in Figure B-28.

Source: Microsoft product screenshots used with permission from Microsoft Corporation

FIGURE B-28   Output of query for average wage rate of employees

Recall the convention that salaried workers are assigned zero dollars an hour. Suppose you want to eliminate the output line for zero dollars an hour because only hourly-rate workers matter for the query. The query setup is shown in Figure B-29.

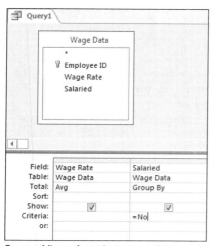

Source: Microsoft product screenshots used with permission from Microsoft Corporation

FIGURE B-29   Query setup for nonsalaried workers only

When you run the query, you will get output for nonsalaried employees only, as shown in Figure B-30.

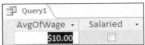

AvgOfWage ▾	Salaried ▾
$10.00	☐

Source: Microsoft product screenshots used with permission from Microsoft Corporation

FIGURE B-30   Query output for nonsalaried workers only

Thus, it is possible to use Criteria in a Totals query, just as you would with a "regular" query. You do not need to print or save the query output, so return to Design view and close the query.

AT THE KEYBOARD

Assume that you want to see two pieces of information for hourly workers: (1) the average wage rate, which you will call Average Rate in the output; and (2) 110 percent of the average rate, which you will call the Increased Rate. To get this information, you can make a calculated field in a new query from a Totals query. In other words, you use one query as a basis for another query.

Create the first query; you already know how to perform certain tasks for this query. The revised heading for the average rate will be Average Rate, so type *Average Rate: Wage Rate* in the Field cell. Note that you want the average of this field. Also, the grouping will be by the Salaried field. (To get hourly workers only, enter *Criteria: No*.) Confirm that your query resembles that in Figure B-31, then save the query and close it.

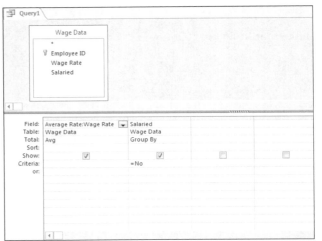

Source: Microsoft product screenshots used with permission from Microsoft Corporation

FIGURE B-31   A totals query with average

Now begin a new query. However, instead of bringing in a table to the query design, select a query. To start a new query, click the Create tab and then click the Query Design button in the Queries group. The Show Table window appears. Click the Queries tab instead of using the default Tables tab, and select the query you just saved as a basis for the new query. The most difficult part of this query is to construct the expression for the calculated field. Conceptually, it is as follows:

Increased Rate: 1.1 * [The current average]

You use the new field name in the new query as the current average, and you treat the new name like a new field:

Increased Rate: 1.1 * [Average Rate]

The query within a query is shown in Figure B-32.

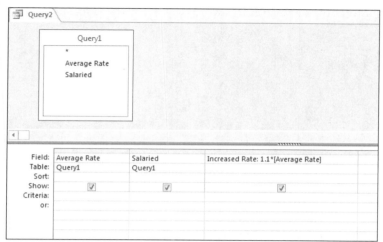

Source: Microsoft product screenshots used with permission from Microsoft Corporation

FIGURE B-32   A query within a query

Figure B-33 shows the output of the new query. Note that the calculated field is formatted.

Source: Microsoft product screenshots used with permission from Microsoft Corporation

FIGURE B-33   Output of an Expression in a Totals query

You do not need to print or save the query output, so return to Design view and close the query.

Using the Date() Function in Queries

Access has two important date function features:

- The built-in Date() function gives you today's date. You can use the function in query criteria or in a calculated field. The function "returns" the day on which the query is run; in other words, it inserts the value where the Date() function appears in an expression.
- Date arithmetic lets you subtract one date from another to obtain the difference—in number of days—between two calendar dates. For example, suppose you create the following expression:
  10/9/2012 – 10/4/2012
  Access would evaluate the expression as the integer 5 (9 minus 4 is 5).

As another example of how date arithmetic works, suppose you want to give each employee a one-dollar bonus for each day the employee has worked. You would need to calculate the number of days between the employee's date of hire and the day the query is run, and then multiply that number by $1.

You would find the number of elapsed days by using the following equation:

Date() – [Date Hired]

Also suppose that for each employee, you want to see the last name, employee ID, and bonus amount. You would set up the query as shown in Figure B-34.

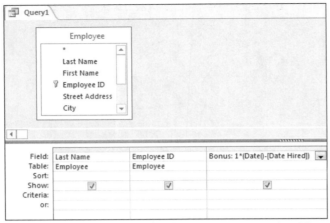

Source: Microsoft product screenshots used with permission from Microsoft Corporation

FIGURE B-34   Date arithmetic in a query

Assume that you set the format of the Bonus field to Currency. The output will be similar to that in Figure B-35, although your Bonus data will be different because you used a different date.

Last Name	Employee ID	Bonus
Brady	09911	$0.00
Howard	11411	$103.00
Johnson	12345	$6,677.00
Smith	14890	$9,921.00
Jones	22282	$3,711.00
Ruth	71460	$5,507.00

Source: Microsoft product screenshots used with permission from Microsoft Corporation

FIGURE B-35   Output of query with date arithmetic

Using Time Arithmetic in Queries

Access also allows you to subtract the values of time fields to get an elapsed time. Assume that your database has a Job Assignments table showing the times that nonsalaried employees were at work during a day. The definition is shown in Figure B-36.

Field Name	Data Type
Employee ID	Short Text
ClockIn	Date/Time
ClockOut	Date/Time
DateWorked	Date/Time

Source: Microsoft product screenshots used with permission from Microsoft Corporation

FIGURE B-36   Date/Time data definition in the Job Assignments table

Assume that the DateWorked field is formatted for Long Date and that the ClockIn and ClockOut fields are formatted for Medium Time. Also assume that for a particular day, nonsalaried workers were scheduled as shown in Figure B-37.

Employee ID	ClockIn	ClockOut	DateWorked	Click to Add
09911	8:30 AM	4:30 PM	Tuesday, September 29, 2015	
11411	9:00 AM	3:00 PM	Tuesday, September 29, 2015	
14890	7:00 AM	5:00 PM	Tuesday, September 29, 2015	

Source: Microsoft product screenshots used with permission from Microsoft Corporation

FIGURE B-37   Display of date and time in a table

You want a query showing the elapsed time that your employees were on the premises for the day. When you add the tables, your screen may show the links differently. Click and drag the Job Assignments, Employee, and Wage Data table icons to look like those in Figure B-38.

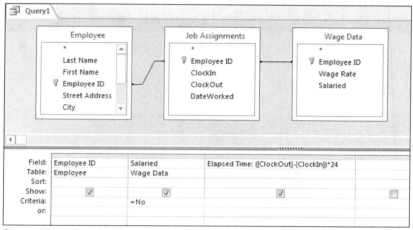

Source: Microsoft product screenshots used with permission from Microsoft Corporation

FIGURE B-38   Query setup for time arithmetic

Figure B-39 shows the output, which looks correct. For example, employee 09911 was at work from 8:30 a.m. to 4:30 p.m., which is eight hours. But how does the odd expression that follows yield the correct answers?

Source: Microsoft product screenshots used with permission from Microsoft Corporation

FIGURE B-39   Query output for time arithmetic

([ClockOut] – [ClockIn]) * 24

Why wouldn't the following expression work?

[ClockOut] – [ClockIn]

Here is the answer: In Access, subtracting one time from the other yields the *decimal* portion of a 24-hour day. Returning to the example, you can see that employee 09911 worked eight hours, which is one-third of a day, so the time arithmetic function yields .3333. That is why you must multiply by 24—to convert from decimals to an hourly basis. Hence, for employee 09911, the expression performs the following calculation: 1/3 × 24 = 8.

Note that parentheses are needed to force Access to do the subtraction *first*, before the multiplication. Without parentheses, multiplication takes precedence over subtraction. For example, consider the following expression:

[ClockOut] – [ClockIn] * 24

In this example, ClockIn would be multiplied by 24, the resulting value would be subtracted from ClockOut, and the output would be a nonsensical decimal number.

Deleting and Updating Queries

The queries presented in this tutorial so far have been Select queries. They select certain data from specific tables based on a given criterion. You also can create queries to update the original data in a database. Businesses use such queries often, and in real time. For example, when you order an item from a Web site, the company's database is updated to reflect your purchase through the deletion of that item from the company's inventory.

Consider an example. Suppose you want to give all nonsalaried workers a $0.50 per hour pay raise. Because you have only three nonsalaried workers, it would be easy to change the Wage Rate data in the table. However, if you had 3,000 nonsalaried employees, it would be much faster and more accurate to change the Wage Rate data by using an Update query that adds $0.50 to each nonsalaried employee's wage rate.

AT THE KEYBOARD

Now you will change each of the nonsalaried employees' pay via an Update query. Figure B-40 shows how to set up the query.

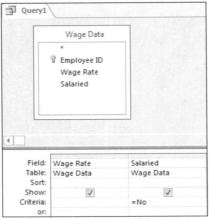

Source: Microsoft product screenshots used with permission from Microsoft Corporation

FIGURE B-40   Query setup for an Update query

So far, this query is just a Select query. Click the Update button in the Query Type group, as shown in Figure B-41.

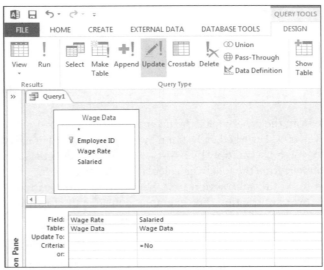

Source: Microsoft product screenshots used with permission from Microsoft Corporation

FIGURE B-41   Selecting a query type

Notice that you now have another line on the QBE grid called Update To:, which is where you specify the change or update the data. Notice that you will update only the nonsalaried workers by using a filter under the Salaried field. Update the Wage Rate data to Wage Rate plus $0.50, as shown in Figure B-42. Note that the update involves the use of brackets [ ], as in a calculated field.

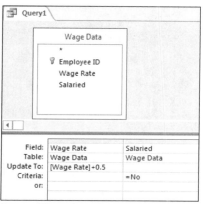

Source: Microsoft product screenshots used with permission from Microsoft Corporation

FIGURE B-42   Updating the wage rate for nonsalaried workers

Now run the query by clicking the Run button in the Results group. If you cannot run the query because it is blocked by Disabled Mode, click the Enable Content button on the Security Warning message bar. When you successfully run the query, the warning message in Figure B-43 appears.

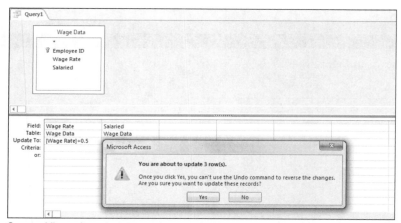

Source: Microsoft product screenshots used with permission from Microsoft Corporation

FIGURE B-43   Update query warning

When you click Yes, the records are updated. Check the updated records by viewing the Wage Data table. Each nonsalaried wage rate should be increased by $0.50. You could add or subtract data from another table as well. If you do, remember to put the field name in square brackets.

Another type of query is the Delete query, which works like Update queries. For example, assume that your company has been purchased by the state of Delaware, which has a policy of employing only state residents. Thus, you must delete (or fire) all employees who are not exclusively Delaware residents. To do that, you would create a Select query. Using the Employee table, you would click the Delete button in the Query Type group, then bring down the State field and filter only those records that were not in Delaware (DE). Do not perform the operation, but note that if you did, the setup would look like the one in Figure B-44.

Source: Microsoft product screenshots used with permission from Microsoft Corporation

FIGURE B-44   Deleting all employees who are not Delaware residents

Using Parameter Queries

A **Parameter query** is actually a type of Select query. For example, suppose your company has 5,000 employees and you want to query the database to find the same kind of information repeatedly, but about different employees each time. For example, you might want to know how many hours a particular employee has worked. You could run a query that you created and stored previously, but run it only for a particular employee.

AT THE KEYBOARD

Create a Select query with the format shown in Figure B-45.

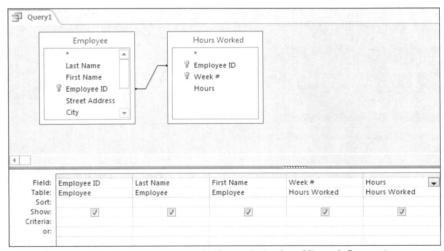

Source: Microsoft product screenshots used with permission from Microsoft Corporation

FIGURE B-45   Design of a Parameter query beginning as a Select query

In the Criteria line of the QBE grid for the Employee ID field, type what is shown in Figure B-46.

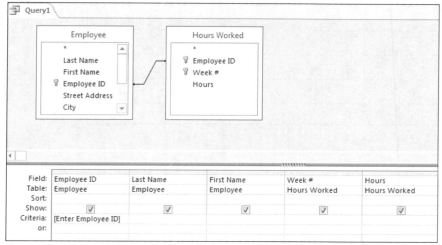

Source: Microsoft product screenshots used with permission from Microsoft Corporation

FIGURE B-46   Design of a Parameter query, continued

Note that the Criteria line uses square brackets, as you would expect to see in a calculated field.
Now run the query. You will be prompted for the employee's ID number, as shown in Figure B-47.

Source: Microsoft product screenshots used with permission from Microsoft Corporation

FIGURE B-47   Enter Parameter Value window

Enter your own employee ID. Your query output should resemble that in Figure B-48.

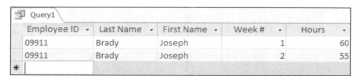

Source: Microsoft product screenshots used with permission from Microsoft Corporation

FIGURE B-48   Output of a Parameter query

MAKING SEVEN PRACTICE QUERIES

This portion of the tutorial gives you additional practice in creating queries. Before making these queries, you must create the specified tables and enter the records shown in the "Creating Tables" section of this tutorial. The output shown for the practice queries is based on those inputs.

AT THE KEYBOARD

For each query that follows, you are given a problem statement and a "scratch area." You also are shown what the query output should look like. Set up each query in Access and then run the query. When you are satisfied with the results, save the query and continue with the next one. Note that you will work with the Employee, Hours Worked, and Wage Data tables.

1. Create a query that shows the employee ID, last name, state, and date hired for employees who live in Delaware *and* were hired after 12/31/99. Perform an ascending sort by employee ID. First click the Sort cell of the field, and then choose Ascending or Descending. Before creating your query, use the table shown in Figure B-49 to work out your QBE grid on paper.

Field					
Table					
Sort					
Show					
Criteria					
Or:					

Source: © 2015 Cengage Learning®

FIGURE B-49 QBE grid template

Your output should resemble that in Figure B-50.

Query 1

Employee ID	Last Name	State	Date Hired
11411	Howard	DE	6/1/2015
22282	Jones	DE	7/15/2004
*			

Source: Microsoft product screenshots used with permission from Microsoft Corporation

FIGURE B-50 Number 1 query output

2. Create a query that shows the last name, first name, date hired, and state for employees who live in Delaware *or* were hired after 12/31/99. The primary sort (ascending) is on last name, and the secondary sort (ascending) is on first name. The Primary Sort field must be to the left of the Secondary Sort field in the query setup. Before creating your query, use the table shown in Figure B-51 to work out your QBE grid on paper.

Field					
Table					
Sort					
Show					
Criteria					
Or:					

Source: © 2015 Cengage Learning®

FIGURE B-51 QBE grid template

If your name was Joseph Brady, your output would look like that in Figure B-52.

Query 2

Last Name	First Name	Date Hired	State
Brady	Joseph	9/15/2015	MD
Howard	Jane	6/1/2015	DE
Johnson	John	6/1/1996	DE
Jones	Sue	7/15/2004	DE
Smith	Albert	7/15/1987	DE
*			

Source: Microsoft product screenshots used with permission from Microsoft Corporation

FIGURE B-52 Number 2 query output

3. Create a query that sums the number of hours worked by U.S. citizens and the number of hours worked by non-U.S. citizens. In other words, create two sums, grouped on citizenship. The heading for total hours worked should be Total Hours Worked. Before creating your query, use the table shown in Figure B-53 to work out your QBE grid on paper.

Field					
Table					
Total					
Sort					
Show					
Criteria					
Or:					

Source: © 2015 Cengage Learning®

FIGURE B-53 QBE grid template

Your output should resemble that in Figure B-54.

Practice Query 3

Total Hours Worked	US Citizen
363	☑
160	☐

Source: Microsoft product screenshots used with permission from Microsoft Corporation

FIGURE B-54 Number 3 query output

4. Create a query that shows the wages owed to hourly workers for Week 1. The heading for the wages owed should be Total Owed. The output headings should be Last Name, Employee ID, Week #, and Total Owed. Before creating your query, use the table shown in Figure B-55 to work out your QBE grid on paper.

Field					
Table					
Sort					
Show					
Criteria					
Or:					

Source: © 2015 Cengage Learning®

FIGURE B-55 QBE grid template

If your name was Joseph Brady, your output would look like that in Figure B-56.

Practice Query 4

Last Name	Employee ID	Week #	Total Owed
Howard	11411	1	$420.00
Smith	14890	1	$475.00
Brady	09911	1	$510.00
*			

Source: Microsoft product screenshots used with permission from Microsoft Corporation

FIGURE B-56 Number 4 query output

5. Create a query that shows the last name, employee ID, hours worked, and overtime amount owed for hourly employees who earned overtime during Week 2. Overtime is paid at 1.5 times the normal hourly rate for all hours worked over 40. Note that the amount shown in the query should be just the overtime portion of the wages paid. Also, this is not a Totals query—amounts should be shown for individual workers. Before creating your query, use the table shown in Figure B-57 to work out your QBE grid on paper.

Field					
Table					
Sort					
Show					
Criteria					
Or:					

Source: © 2015 Cengage Learning®

FIGURE B-57   QBE grid template

If your name was Joseph Brady, your output would look like that in Figure B-58.

Practice Query 5			
Last Name	Employee ID	Hours	OT Pay
Howard	11411	50	$157.50
Brady	09911	55	$191.25

Source: Microsoft product screenshots used with permission from Microsoft Corporation

FIGURE B-58   Number 5 query output

6. Create a Parameter query that shows the hours employees have worked. Have the Parameter query prompt for the week number. The output headings should be Last Name, First Name, Week #, and Hours. This query is for nonsalaried workers only. Before creating your query, use the table shown in Figure B-59 to work out your QBE grid on paper.

Field					
Table					
Sort					
Show					
Criteria					
Or:					

Source: © 2015 Cengage Learning®

FIGURE B-59   QBE grid template

Run the query and enter 2 when prompted for the week number. Your output should look like that in Figure B-60.

Practice Query 6			
Last Name	First Name	Week #	Hours
Howard	Jane	2	50
Smith	Albert	2	40
Brady	Joseph	2	55

Source: Microsoft product screenshots used with permission from Microsoft Corporation

FIGURE B-60   Number 6 query output

7. Create an Update query that gives certain workers a merit raise. First, you must create an additional table, as shown in Figure B-61.

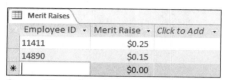

Merit Raises		
Employee ID ▾	Merit Raise ▾	Click to Add ▾
11411	$0.25	
14890	$0.15	
*	$0.00	

Source: Microsoft product screenshots used with permission from Microsoft Corporation

FIGURE B-61   Merit Raises table

Create a query that adds the Merit Raise to the current Wage Rate for employees who will receive a raise. When you run the query, you should be prompted with *You are about to update two rows.* Check the original Wage Data table to confirm the update. Before creating your query, use the table shown in Figure B-62 to work out your QBE grid on paper.

Field					
Table					
Update to					
Criteria					
Or:					

Source: © 2015 Cengage Learning®

FIGURE B-62   QBE grid template

CREATING REPORTS

Database packages let you make attractive management reports from a table's records or from a query's output. If you are making a report from a table, the Access report generator looks up the data in the table and puts it into report format. If you are making a report from a query's output, Access runs the query in the background (you do not control it or see it happen) and then puts the output in report format.

There are different ways to make a report. One method is to create one from scratch in Design view, but this tedious process is not explained in this tutorial. A simpler way is to select the query or table on which the report is based and then click Report on the Create tab. This streamlined method of creating reports is explained in this tutorial.

Creating a Grouped Report

This tutorial assumes that you already know how to create a basic ungrouped report, so this section teaches you how to make a grouped report. If you do not know how to create an ungrouped report, you can learn by following the first example in the upcoming section.

AT THE KEYBOARD

Suppose you want to create a report from the Hours Worked table. Select the table by clicking it once. Click the Create tab, then click Report in the Reports group. A report appears, as shown in Figure B-63.

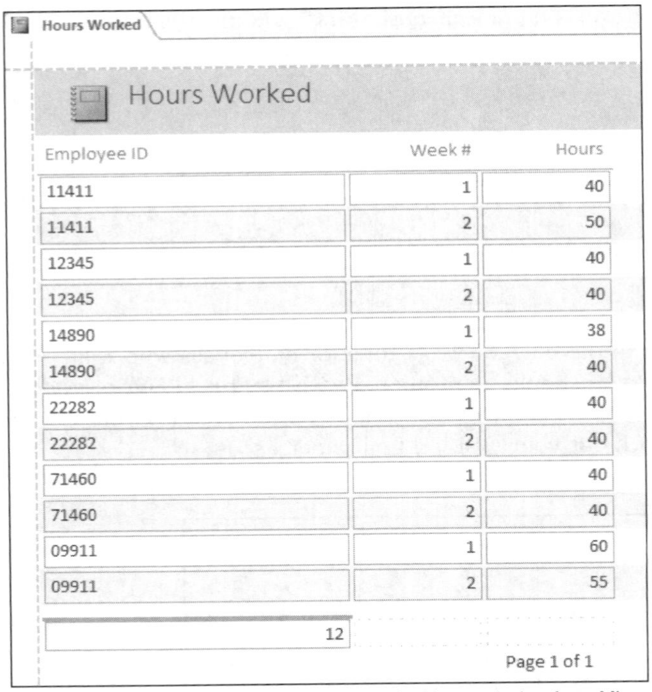

Source: Microsoft product screenshots used with permission from Microsoft Corporation

FIGURE B-63   Initial report based on a table

On the Design tab, select the Group & Sort button in the Grouping & Totals group. Your report will have an additional selection at the bottom, as shown in Figure B-64.

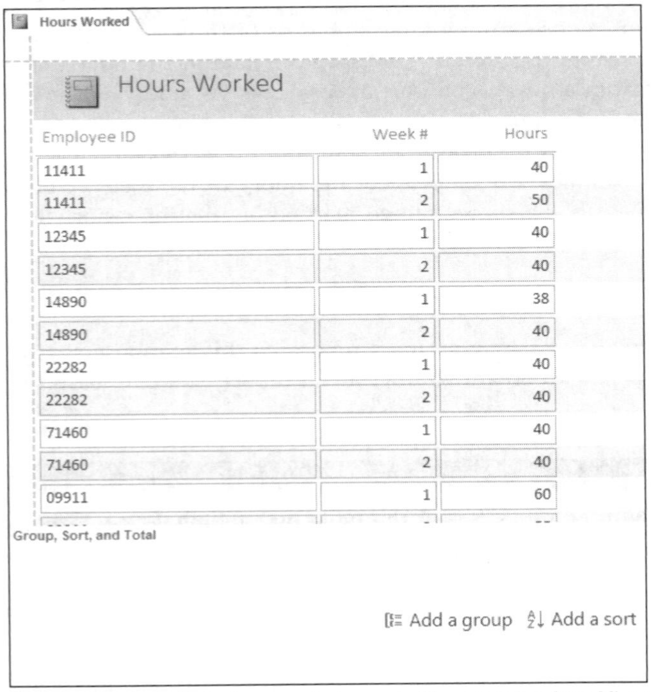

Source: Microsoft product screenshots used with permission from Microsoft Corporation

FIGURE B-64   Report with grouping and sorting options

Click the Add a group button at the bottom of the report, and then select Employee ID. Your report will be grouped as shown in Figure B-65.

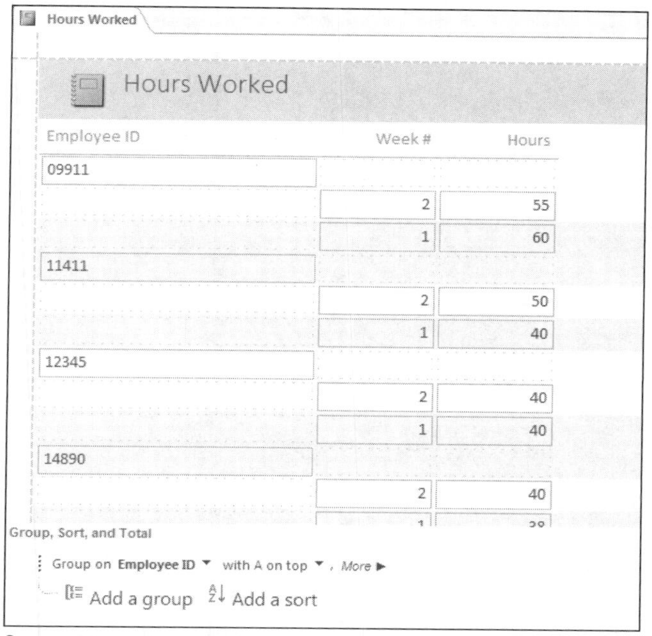

Source: Microsoft product screenshots used with permission from Microsoft Corporation

FIGURE B-65   Grouped report

To complete this report, you need to total the hours for each employee by selecting the Hours column heading. Your report will show that the entire column is selected. On the Design tab, click the Totals button in the Grouping & Totals group, and then choose Sum from the menu, as shown in Figure B-66.

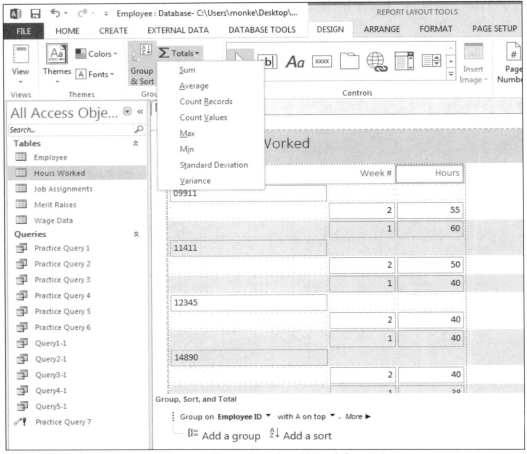

Source: Microsoft product screenshots used with permission from Microsoft Corporation

FIGURE B-66   Totaling the hours

Your report will look like the one in Figure B-67.

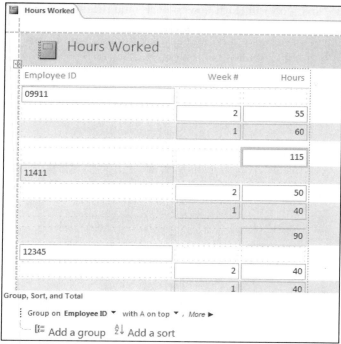

Source: Microsoft product screenshots used with permission from Microsoft Corporation

FIGURE B-67   Completed report

Your report is currently in Layout view. To see how the final report looks when printed, click the Design tab and select Report View from the Views group. Your report looks like the one in Figure B-68, although only a portion is shown in the figure.

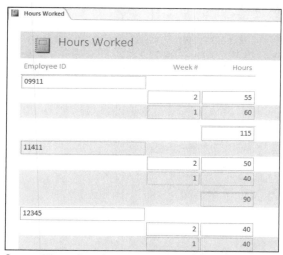

Source: Microsoft product screenshots used with permission from Microsoft Corporation

FIGURE B-68   Report in Report view

NOTE

To change the picture or logo in the upper-left corner of the report when in Layout view, click the notebook symbol and press the Delete key. You can insert a logo in place of the notebook by clicking the Design tab and then clicking the Insert Image button in the Controls group.

Moving Fields in Layout View

If you group records based on more than one field in a report, the report will have an odd "staircase" look or display repeated data, or it will have both problems. Next, you will learn how to overcome these problems in Layout view.

Suppose you make a query that shows an employee's last name, first name, week number, and hours worked, and then you make a report from that query, grouping on last name only. See Figure B-69.

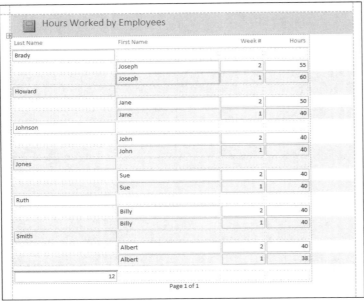

Source: Microsoft product screenshots used with permission from Microsoft Corporation

FIGURE B-69 Query-based report grouped on last name

As you preview the report, notice the repeating data from the First Name field. In the report shown in Figure B-69, notice that the first name repeats for each week worked—hence, the staircase effect. The Week # and Hours fields are shown as subordinate to Last Name, as desired.

Suppose you want the last name and first name to appear on the same line. If so, take the report into Layout view for editing. Click the first record for the First Name (in this case, Joseph), and drag the name up to the same line as the Last Name (in this case, Brady). Your report will now show the First Name on the same line as Last Name, thereby eliminating the staircase look, as shown in Figure B-70.

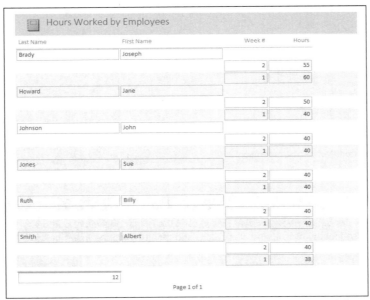

Source: Microsoft product screenshots used with permission from Microsoft Corporation

FIGURE B-70 Report in Layout view with Last Name and First Name on the same line

You can now add the sum of Hours for each group. Also, if you want to add more fields such as Street Address and Zip, you can repeat the preceding procedure.

IMPORTING DATA

Text or spreadsheet data is easy to import into Access. In business, it is often necessary to import because companies use disparate systems. For example, assume that your healthcare coverage data i human resources manager's computer in a Microsoft Excel spreadsheet. Open the Excel application and then create a spreadsheet using the data shown in Figure B-71.

	A	B	C
1	Employee ID	Provider	Level
2	11411	BlueCross	family
3	12345	BlueCross	family
4	14890	Coventry	spouse
5	22282	None	none
6	71460	Coventry	single
7	Your ID	BlueCross	single

Source: Microsoft product screenshots used with permission from Microsoft Corporation

FIGURE B-71   Excel data

Save the file and then close it. Now you can easily import the spreadsheet data into a new table in Access. With your Employee database open, click the External Data tab, then click Excel in the Import & Link group. Browse to find the Excel file you just created, and make sure the first radio button is selected to import the source data into a new table in the current database (see Figure B-72). Click OK.

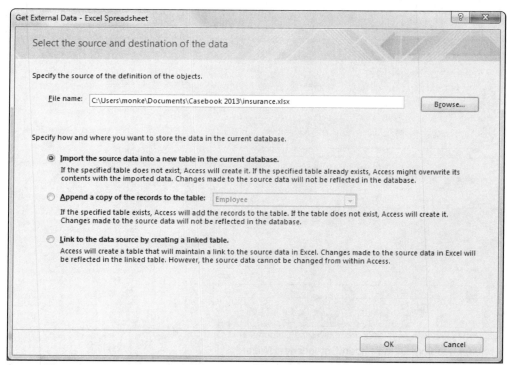

Source: Microsoft product screenshots used with permission from Microsoft Corporation

FIGURE B-72   Importing Excel data into a new table

Choose the correct worksheet. Your next screen should look like the one in Figure B-73.

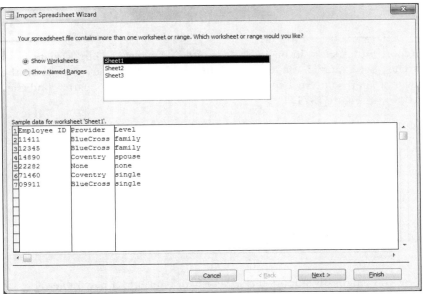

FIGURE B-73   First screen in the Import Spreadsheet Wizard

Choose Next, and then make sure to select the First Row Contains Column Headings box, as shown in Figure B-74.

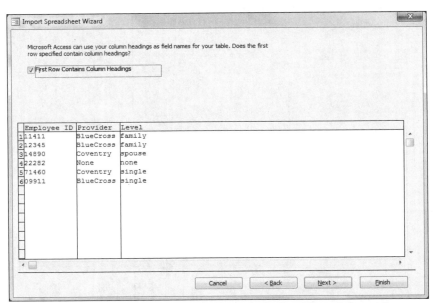

FIGURE B-74   Choosing column headings in the Import Spreadsheet Wizard

Choose Next. Accept the default setting for each field you are importing on the screen. Each field is assigned a text data type, which is correct for this table. Your screen should look like the one in Figure B-75.

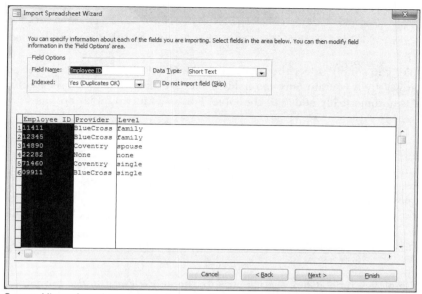

Source: Microsoft product screenshots used with permission from Microsoft Corporation

FIGURE B-75   Choosing the data type for each field in the Import Spreadsheet Wizard

Choose Next. In the next screen of the wizard, you will be prompted to create an index—that is, to define a primary key. Because you will store your data in a new table, choose your own primary key (Employee ID), as shown in Figure B-76.

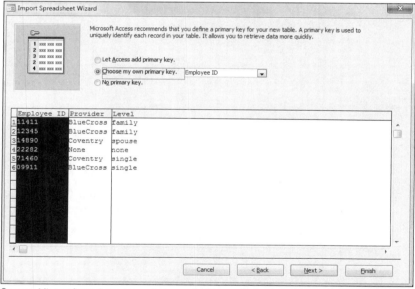

Source: Microsoft product screenshots used with permission from Microsoft Corporation

FIGURE B-76   Choosing a primary key field in the Import Spreadsheet Wizard

Continue through the wizard, giving your table an appropriate name. After importing the table, take a look at its design by right-clicking the table and choosing Design View. Note that each field is very wide. Adjust the field properties as needed.

MAKING FORMS

Forms simplify the process of adding new records to a table. Creating forms is easy, and they can be applied to one or more tables.

When you base a form on one table, you simply select the table, click the Create tab, and then select Form from the Forms group. The form will then contain only the fields from that table. When data is entered into the form, a complete new record is automatically added to the table. Forms with two tables are discussed next.

Making Forms with Subforms

You also can create a form that contains a subform, which can be useful when the form is based on two or more tables. Return to the example Employee database to see how forms and subforms would be useful for viewing all of the hours that each employee worked each week. Suppose you want to show all of the fields from the Employee table; you also want to show the hours each employee worked by including all fields from the Hours Worked table as well.

To create the form and subform, first create a simple one-table form on the Employee table. Follow these steps:

1. Click once to select the Employee table. Click the Create tab, then click Form in the Forms group. After the main form is complete, it should resemble the one in Figure B-77.

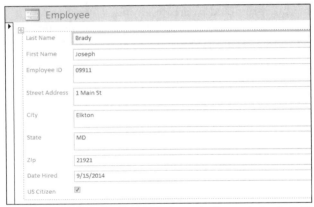

Source: Microsoft product screenshots used with permission from Microsoft Corporation

FIGURE B-77   The Employee form

2. To add the subform, take the form into Design view. On the Design tab, make sure that the Use Control Wizards option is selected, scroll to the bottom row of buttons in the Controls group, and click the Subform/Subreport button, as shown in Figure B-78.

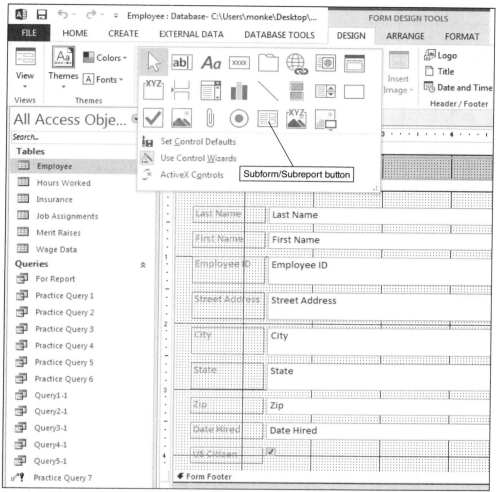

Source: Microsoft product screenshots used with permission from Microsoft Corporation

FIGURE B-78 The Subform/Subreport button

3. Use your cursor to stretch out the box under your main form. You might need to expand the area beneath the main form. The window shown in Figure B-79 appears.

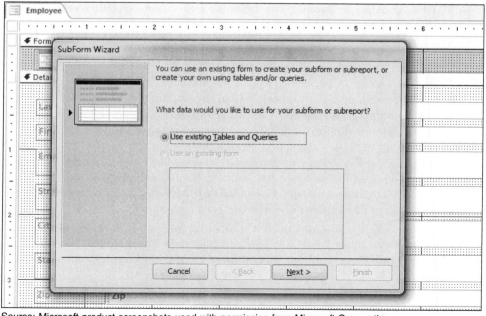

Source: Microsoft product screenshots used with permission from Microsoft Corporation

FIGURE B-79 Adding a subform

4. Select Use existing Tables and Queries, click Next, and then select Table: Hours Worked from the Tables/Queries drop-down list. Select all available fields. Click Next, select Choose from a list, click Next again, and then click Finish. Select the Form view. Your form and subform should resemble Figure B-80. You may need to stretch out the subform box in Design view if all fields are not visible.

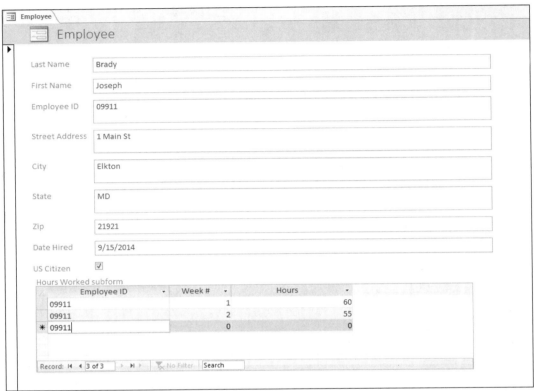

Source: Microsoft product screenshots used with permission from Microsoft Corporation

FIGURE B-80 Form with subform

TROUBLESHOOTING COMMON PROBLEMS

Access is a powerful program, but it is complex and sometimes difficult for new users. People sometimes unintentionally create databases that have problems. Some of these common problems are described below, along with their causes and corrections.

1. *"I saved my database file, but I can't find it on my computer or my external secondary storage medium! Where is it?"*

    You saved your file to a fixed disk or a location other than the Documents folder. Use the Windows Search option to find all files ending in .accdb (search for *.accdb). If you saved the file, it is on the hard drive (C:\) or a network drive. Your site assistant can tell you the drive designators.

2. *"What is a 'duplicate key field value'? I'm trying to enter records into my Sales table. The first record was for a sale of product X to customer 101, and I was able to enter that one. But when I try to enter a second sale for customer #101, Access tells me I already have a record with that key field value. Am I allowed to enter only one sale per customer?"*

    Your primary key field needs work. You may need a compound primary key—a combination of the customer number and some other field(s). In this case, the customer number, product number, and date of sale might provide a unique combination of values, or you might consider using an invoice number field as a key.

3. *"My query reads 'Enter Parameter Value' when I run it. What is that?"*

This problem almost always indicates that you have misspelled a field name in an expression in a Criteria field or calculated field. Access is very fussy about spelling; for example, it is case-sensitive. Access is also "space-sensitive," meaning that when you insert a space in a field name when defining a table, you must also include a space in the field name when you reference it in a query expression. Fix the typo in the query expression.

4. *"I'm getting an enormous number of rows in my query output—many times more than I need. Most of the rows are duplicates!"*

This problem is usually caused by a failure to link all of the tables you brought into the top half of the query generator. The solution is to use the manual click-and-drag method to link the common fields between tables. The spelling of the field names is irrelevant because the link fields need not have the same spelling.

5. *"For the most part, my query output is what I expected, but I am getting one or two duplicate rows or not enough rows."*

You may have linked too many fields between tables. Usually, only a single link is needed between two tables. It is unnecessary to link each common field in all combinations of tables; it is usually sufficient to link the primary keys. A simplistic explanation for why overlinking causes problems is that it causes Access to "overthink" and repeat itself in its answer.

On the other hand, you might be using too many tables in the query design. For example, you brought in a table, linked it on a common field with some other table, but then did not use the table. In other words, you brought down none of its fields, and/or you used none of its fields in query expressions. In this case, if you got rid of the table, the query would still work. Click the unneeded table's header at the top of the QBE area, and press the Delete key to see if you can make the few duplicate rows disappear.

6. *"I expected six rows in my query output, but I got only five. What happened to the other one?"*

Usually, this problem indicates a data entry error in your tables. When you link the proper tables and fields to make the query, remember that the linking operation joins records from the tables *on common values* (*equal* values in the two tables). For example, if a primary key in one table has the value "123," the primary key or the linking field in the other table should be the same to allow linking. Note that the text string "123" is not the same as the text string " 123"—the space in the second string is considered a character too. Access does not see unequal values as an error. Instead, Access moves on to consider the rest of the records in the table for linking. The solution is to examine the values entered into the linked fields in each table and fix any data entry errors.

7. *"I linked fields correctly in a query, but I'm getting the empty set in the output. All I get are the field name headings!"*

You probably have zero common (equal) values in the linked fields. For example, suppose you are linking on Part Number, which you declared as text. In one field, you have part numbers "001," "002," and "003"; in the other table, you have part numbers "0001," "0002," and "0003." Your tables have no common values, which means that no records are selected for output. You must change the values in one of the tables.

8. *"I'm trying to count the number of today's sales orders. A Totals query is called for. Sales are denoted by an invoice number, and I made that a text field in the table design. However, when I ask the Totals query to 'Sum' the number of invoice numbers, Access tells me I cannot add them up! What is the problem?"*

Text variables are words! You cannot add words, but you can count them. Use the Count Totals operator (not the Sum operator) to count the number of sales, each being denoted by an invoice number.

9. *"I'm doing time arithmetic in a calculated field expression. I subtracted the Time In from the Time Out and got a decimal number! I expected eight hours, and I got the number .33333. Why?"*

[Time Out] – [Time In] yields the decimal percentage of a 24-hour day. In your case, eight hours is one-third of a day. You must complete the expression by multiplying by 24: ([Time Out] – [Time In]) * 24. Don't forget the parentheses.

10. *"I formatted a calculated field for Currency in the query generator, and the values did show as currency in the query output; however, the report based on the query output does not show the dollar sign in its output. What happened?"*

Go to the report Design view. A box in one of the panels represents the calculated field's value. Click the box and drag to widen it. That should give Access enough room to show the dollar sign as well as the number in the output.

11. *"I told the Report Wizard to fit all of my output to one page. It does print to one page, but some of the data is missing. What happened?"*

Access fits all the output on one page by leaving data out. If you can tolerate having the output on more than one page, deselect the Fit to a Page option in the wizard. One way to tighten output is to enter Design view and remove space from each box that represents output values and labels. Access usually provides more space than needed.

12. *"I grouped on three fields in the Report Wizard, and the wizard prints the output in a staircase fashion. I want the grouping fields to be on one line. How can I do that?"*

Make adjustments in Design view and Layout view. See the "Creating Reports" section of this tutorial for instructions on making these adjustments.

13. *"When I create an Update query, Access tells me that zero rows are updating or more rows are updating than I want. What is wrong?"*

If your Update query is not set up correctly (for example, if the tables are not joined properly), Access will either try not to update anything, or it will update all of the records. Check the query, make corrections, and run it again.

14. *"I made a Totals query with a Sum in the Group By row and saved the query. Now when I go back to it, the Sum field reads 'Expression,' and 'Sum' is entered in the field name box. Is that wrong?"*

Access sometimes changes the Sum field when the query is saved. The data remains the same, and you can be assured your query is correct.

15. *"I cannot run my Update query, but I know it is set up correctly. What is wrong?"*

Check that you have clicked the Enable Content button on the Security Warning message bar.

PRELIMINARY CASE: THE TEXTBOOK BARTERING DATABASE

Setting Up a Relational Database to Create Tables, Forms, Queries, and Reports

PREVIEW

In this case, you will create a relational database for a textbook bartering service. First, you will create four tables and populate them with data. Next, you will create a form and subform for recording books and their orders. You will create five queries: a select query, a parameter query, an update query, a totals query, and a query used as the basis for a report. Finally, you will create the report from the fifth query.

PREPARATION

- Before attempting this case, you should have some experience using Microsoft Access.
- Complete any part of Tutorial B that your instructor assigns, or refer to the tutorial as necessary.

BACKGROUND

Your roommate is minoring in environmental science and is very interested in keeping landfills as empty as possible. She recently learned about sites such as eBay or Craigslist where users can barter for used goods without exchanging any money. Users either can give the site items to barter and earn credits or they can purchase credits to "buy" goods. You and your roommate have created a similar bartering service for used textbooks on your college campus. Your organization would be the go-between for students to give and take used books for classes. As a prototype for the service, you suggest creating an Access database to track and record all bartering; you have experience using the program from your information systems coursework. Once your prototype is up and running with students using the service, you will approach a local bank and request financing for the service and database.

    Your first tasks are to design the database, create the tables, and populate them with current data. You've decided to begin in a simple fashion, so your database design includes only four tables, as shown in Figures 1-1, 1-2, 1-3, and 1-4:

- Books, which keeps track of book ID numbers, book titles, the number of copies of each book the organization has available for bartering, and the number of credits the book is worth for trading
- Students, which records student information such as name, address, phone numbers, and e-mail address
- Orders, which maintains records for each student's ID number, the date each book was requested, and the ID number of the desired book
- Credits, which keeps track of how many credits each student has for spending on used books

    After the database tables are complete and populated with data, you want to computerize several common tasks. First, you need a streamlined way to know which students are requesting each book. You recognize that a form and subform would be ideal for recording information about book bartering.

To keep track of inventory, you think it might be beneficial to know which books have large amounts in stock so you can prepare for the demands of large classes before the upcoming semester. This listing will be easy to complete using a basic select query that identifies any book for which you have more than 50 copies in stock.

Students would like to know if a particular book required for a class is available for bartering. To satisfy this request, you suggest a parameter query that prompts the user to enter a book name. After a student has requested and obtained a few books, the student's number of credits will have to be updated accordingly. You realize that an update query will accomplish this task.

Your roommate thinks the two of you should keep an eye on how many books are being requested so you can request a bank loan to finance your service and database. She would like to see a list that includes all books requested so far and sorts the titles from greatest number ordered to the least. You are confident that a totals query will satisfy your roommate's request.

Finally, the two of you would like a comprehensive list of students who have registered for your service, along with the books they have requested and obtained. You decide that the best solution is to use a query that feeds into a report. The report will show each student who has used the service, the books each has received, and a count of the total number of credits the books are worth.

ASSIGNMENT 1: CREATING TABLES

Use Microsoft Access to create tables that contain the fields shown in Figures 1-1 through 1-4; you learned about these tables in the Background section. Populate the database tables as shown. Add your name to the Students table with a fictitious ID number; complete the entry by adding your address, phone numbers, and e-mail address. Order yourself two books and give yourself a credit of 50.

This database contains the following four tables:

Books				
Book ID ▾	Book Title ▾	Number of Copies ▾	Credits ▾	Click to Add ▾
101	Problem Solving Cases in MS Access and Excel	10	3	
102	Principles of Accounting I	120	5	
103	General Chemistry	50	7	
104	Physics Made Easy	15	2	
105	Differential Equations	30	5	
106	Engineering Analysis	120	4	
107	Marketing Research	25	7	
108	Operations Research	45	8	
109	Financial Analysis	100	3	
110	Effective Communication	200	2	
*		0	0	

Source: Microsoft product screenshots used with permission from Microsoft Corporation

FIGURE 1-1   The Books table

*add own name

Students										
Student ID ▾	Last Name ▾	First Name ▾	Address ▾	City ▾	State ▾	Zip ▾	Cell Phone ▾	Telephone ▾	Email Address ▾	Click to Add ▾
B-17	Brewster	Angela	9 Pleasant Way	Philadelphia	PA	19101	215-876-3376	215-887-4673	abrew@yahoo.com	
F-59	Freeny	Geraldine	1012 Peachtree St	Philadelphia	PA	19111	215-765-1263	215-887-2342	gf59@gmail.com	
L-29	Lopez	Maria	5490 West 5th	Philadelphia	PA	19101	569-001-0989	215-234-8876	mrl@yahoo.com	
M-62	May	Allie	59 W. Central Ave	Philadelphia	PA	19111	215-887-3829	215-998-3928	belle@comcast.net	
P-91	Pettyjon	Jane	89 Orchard	Philadelphia	PA	19101	215-342-9087	215-887-9238	jp82@comcast.net	
Q-13	Quinn	Samantha	54 Oak Ave	Philadelphia	PA	19125	569-984-3894	215-987-3427	quinn45@gmail.com	
S-63	Signorelli	Patricia	1700 E. Lincoln Ave	Philadelphia	PA	19111	215-121-4736	215-765-3342	patti1@gmail.com	
Z-30	Zecon	Joyce	58 W. Central Ave	Philadelphia	PA	19125	215-776-4536	215-675-0091	zeconj@comcast.net	
*										

Source: Microsoft product screenshots used with permission from Microsoft Corporation

FIGURE 1-2   The Students table

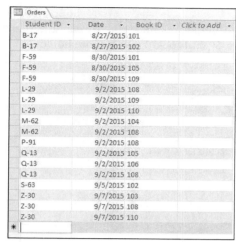

*add own order for 2 books

Cannot create primary key → won't allow duplicates

Source: Microsoft product screenshots used with permission from Microsoft Corporation

FIGURE 1-3   The Orders table

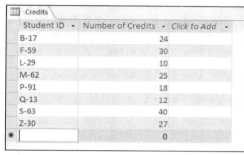

*add credits for 50

Source: Microsoft product screenshots used with permission from Microsoft Corporation

FIGURE 1-4   The Credits table

ASSIGNMENT 2: CREATING A FORM, QUERIES, AND A REPORT

Assignment 2A: Creating a Form

Create a form for easy recording of books and their orders. The main form should be based on the Books table and the subform should include the fields from the Orders table. Save the form as Books. View one record; if required by your instructor, print the record. Your output should resemble that shown in Figure 1-5.

Books

Book ID	101
Book Title	Problem Solving Cases in MS Access and Excel
Number of Copies	10
Credits	3

Orders subform

Student ID	Date	Book ID
B-17	8/27/2015	101
F-59	8/30/2015	101
*		101

Record: 3 of 3   Search

Source: Microsoft product screenshots used with permission from Microsoft Corporation

FIGURE 1-5   The Books form with subform

Assignment 2B: Creating a Select Query

Create a query to list all books that have more than 50 copies in stock. Include columns that display the Book ID, Book Title, and Number of Copies. Save the query as Greater Than 50 In Stock. Your output should resemble that shown in Figure 1-6. Print the output if desired.

Book ID	Book Title	Number of Copies
102	Principles of Accounting I	120
106	Engineering Analysis	120
109	Financial Analysis	100
110	Effective Communication	200
*		0

Source: Microsoft product screenshots used with permission from Microsoft Corporation

FIGURE 1-6　Greater Than 50 In Stock query

Assignment 2C: Creating a Parameter Query

Create a parameter query that prompts for a specific book and then lists the number of copies available in inventory and how many credits the book is worth. The query should include columns for Book ID, Book Title, Number of Copies, and Credits. Save the query as What Book? Your output should resemble Figure 1-7 after you enter "Principles of Accounting I" at the prompt.

Book ID	Book Title	Number of Copies	Credits
102	Principles of Accounting I	120	5
*		0	0

Source: Microsoft product screenshots used with permission from Microsoft Corporation

FIGURE 1-7　What Book? query

Assignment 2D: Creating an Update Query

Create a query that updates students' credits to account for the books they've ordered and received. Click the Run button to test the query. When prompted to change the records, answer "Yes." Save the query as Credit Update.

Assignment 2E: Creating a Totals Query

Create a totals query that lists the titles of all books in inventory and shows how many orders have been received for each. In the output, display columns for Book Title and Number of Books Ordered. Sort the list of books from the most ordered to the least. Note the change to the Number of Books Ordered column heading. Save your query as Books Ordered. Your output should resemble that shown in Figure 1-8. Print the output if desired.

primary key not linked so can't complete

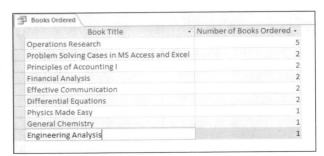

Book Title	Number of Books Ordered
Operations Research	5
Problem Solving Cases in MS Access and Excel	2
Principles of Accounting I	2
Financial Analysis	2
Effective Communication	2
Differential Equations	2
Physics Made Easy	1
General Chemistry	1
Engineering Analysis	1

Source: Microsoft product screenshots used with permission from Microsoft Corporation

FIGURE 1-8　Books Ordered query

Assignment 2F: Generating a Report

✗create query first

Generate a report based on a query that summarizes student orders and credits. The query should display columns for students' Last Name, First Name, and Number of Credits, plus the Book Title and Credits each book is worth. Save the query as Order Report. From that query, create a report that groups each student's activity and totals up the credits for their received books. Make any needed adjustments to the output to avoid repeating names and to ensure that all fields and data are visible. Ensure that "Order Report" appears as the title at the top of the report, and save the report under the same name. Your report output should resemble that shown in Figure 1-9, although only a portion of the report appears in the figure.

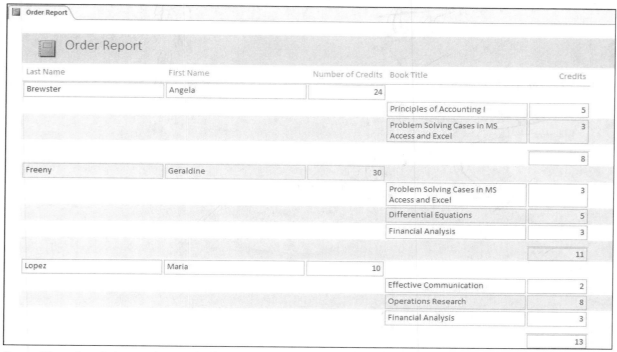

Source: Microsoft product screenshots used with permission from Microsoft Corporation

FIGURE 1-9   Order report

    If you are working with a portable storage disk or USB thumb drive, make sure that you remove it *after* closing the database file.

DELIVERABLES

Assemble the following deliverables for your instructor, either electronically or in printed form:

1. Four tables
2. Form and subform: Books
3. Query 1: Greater Than 50 In Stock
4. Query 2: What Book?
5. Query 3: Credit Update
6. Query 4: Books Ordered
7. Query 5: Order Report
8. Report: Order Report
9. Any other required printouts or electronic media

    Staple all the pages together. Write your name and class number at the top of the page. If required, make sure that your electronic media are labeled.

SMGramley - HW#1 - Solution

BICYCLE REPAIR SHOP DATABASE

Designing a Relational Database to Create Tables, Forms, Queries, and Reports

PREVIEW

In this case, you will design a relational database for a business that repairs bicycles. After your design is completed and correct, you will create database tables and populate them with data. Then you will produce one form with a subform, four queries, and two reports. The queries will address the following questions or tasks: Which customers are under the age of 25? What jobs are performed by a specific technician? Which repair jobs are the most popular? What will repair prices be if the cost of each job is increased by 5 percent? Your reports will summarize the total amount of money owed by each customer, the jobs performed by each technician, and the price of each repair.

PREPARATION

- Before attempting this case, you should have some experience in database design and in using Microsoft Access.
- Complete any part of Tutorial A that your instructor assigns.
- Complete any part of Tutorial B that your instructor assigns, or refer to the tutorial as necessary.
- Refer to Tutorial F as necessary.

BACKGROUND

You are an avid cyclist and enjoy biking with your favorite uncle, Joe. He recently retired from his military job and opened a bicycle repair shop in the town where your university is located. Bicycle repair has been his hobby for many years, and now he wants to try it as a business. You visit his new shop and begin chatting about your information systems course. Uncle Joe finds out that you are proficient with designing and implementing databases, so he hires you over the summer to create a prototype system for his business.

On your first day of work, you sit down with Uncle Joe to learn as much as you can about the business. Joe spends a lot of time talking about his customers, and he keeps track of frequent customers. He knows their full names, addresses, phone numbers, and even their birthdates. He likes to record customer birthdates because he gives a discount to anyone who might be a university student. Joe also explains that he performs five repair services: tune-ups, adjustments to brakes and gears, chain installation, and wheel trueing. Each service costs a different amount of money. Uncle Joe has hired four of his friends as technicians, and he calls them by their first names. He says some of the technicians are left-handed, which is important because some tools are made specifically for right-handed or left-handed workers and he must have the proper tools for the jobs required.

When customers come into the shop with bicycles to be serviced, they most likely need multiple repairs. For example, a customer might want her bicycle to be tuned up, the wheels to be trued, and the gears adjusted. Uncle Joe wants all those jobs included in one receipt or bill. He requests an easy way to record each visit, including a list of the repair jobs performed. You suggest that a form and subform would be an excellent solution to this request.

Although Joe is friendly with his technicians, he runs a tight ship and likes to know how they use their time. For example, he'd like to be able to type in a technician number and see how many jobs that technician has completed. You are confident that a parameter query will fulfill Joe's needs. In keeping with Joe's managerial style, he'd also like to have a total of the number of times each job is done, listed from the most popular job to the least popular.

As mentioned, Uncle Joe is fond of college students and would like a listing of customers under age 25. He also knows he will have to raise all his prices by 5 percent at some point in the future. He opened his business with very low prices to attract customers, but he realizes that his costs are not being covered fully. You explain to Joe that it's easy to increase all repair prices with an update query.

After your initial discussion with Uncle Joe, you begin to think of additional features that he would find useful. The first new feature you suggest is a customer report. You can create a neatly formatted report that lists each customer, his or her address, the repair date, the repair job, and its price. For each customer, you can have the report display the total amount of money paid to the business.

In a similar vein, you suggest a report that describes each technician's work. You can list each set of jobs performed by each technician, the date and price of each repair job, and the total amount of money charged for each technician's work. Joe agrees that these reports fit well with his management style and business, and he encourages you to begin the project.

ASSIGNMENT 1: CREATING THE DATABASE DESIGN

In this assignment, you design your database tables using a word-processing program. Pay close attention to the logic and structure of the tables. Do not start developing your Access database in Assignment 2 before getting feedback from your instructor on Assignment 1. Keep in mind that you need to examine the requirements in Assignment 2 to design your fields and tables properly. It is good programming practice to look at the required outputs before beginning your design. When designing the database, observe the following guidelines:

- First, determine the tables you will need by listing the name of each table and the fields it should contain. Avoid data redundancy. Do not create a field if it can be created by a calculated field in a query.
- You will need transaction tables. Think about the business events that occur with each customer's actions. Keep in mind that most customers have more than one job performed during each visit. Avoid duplicating data.
- Document your tables using the table feature of your word processor. Your tables should resemble the format shown in Figure 2-1.
- You must mark the appropriate key field(s) by entering an asterisk (*) next to the field name. Keep in mind that some tables might need a compound primary key to uniquely identify a record within a table.
- Print the database design.

Table Name	
Field Name	**Data Type (text, numeric, currency, etc.)**
…	…
…	…

Source: © 2015 Cengage Learning®

FIGURE 2-1   Table design

NOTE
Have your design approved before beginning Assignment 2; otherwise, you may need to redo Assignment 2.

ASSIGNMENT 2: CREATING THE DATABASE, QUERIES, AND REPORTS

In this assignment, you first create database tables in Access and populate them with data. Next, you create a form, four queries, and two reports.

Assignment 2A: Creating Tables in Access

In this part of the assignment, you create your tables in Access. Use the following guidelines:

- Enter at least 10 customer records, and make sure that some customers are younger than 25 years old. Add yourself as a customer.
- Make sure that each customer visits the bicycle repair shop at least once and has two or more jobs performed during each visit.

- The shop offers five different types of bicycle repair jobs, as outlined previously. Make sure each job costs a different amount of money.
- Create records for four technicians and assign them to the various customers' jobs.
- Appropriately limit the size of the text fields; for example, a telephone number does not need the default length of 255 characters.
- Print all tables if your instructor requires it.

Assignment 2B: Creating Forms, Queries, and Reports

You must generate one form with a subform, four queries, and two reports, as outlined in the Background section of this case.

Form

Create a form and subform based on your Visits table and Visit Line Item table (or whatever you named these tables). Save the form as Visits. Your form should resemble that shown in Figure 2-2.

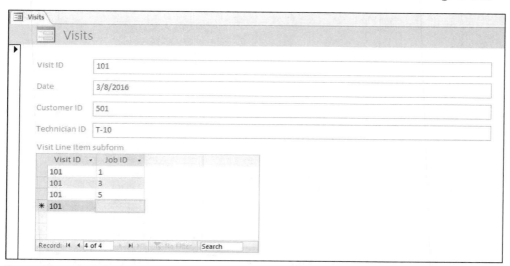

Source: Microsoft product screenshots used with permission from Microsoft Corporation

FIGURE 2-2 Visits form and subform

Query 1

Create a parameter query called Jobs by Technician Number that prompts for a technician number and then displays columns for Job Description and Number of Times the technician has performed each job. Do not show the technician number in the output. Note that Number of Times is a change in column heading from the default setting in the query generator. In Figure 2-3, the user entered T-13 as the technician number. Your output should resemble that shown in Figure 2-3, although your data will be different.

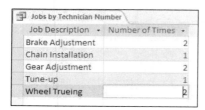

Source: Microsoft product screenshots used with permission from Microsoft Corporation

FIGURE 2-3 Jobs by Technician Number query

Query 2

Create a query called Customers Under 25 that displays columns for the Last Name, First Name, Street Address, City, State, and Zip code of customers who are less than 25 years old. Your output should look like that shown in Figure 2-4, although your data will be different.

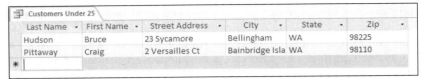

Source: Microsoft product screenshots used with permission from Microsoft Corporation

FIGURE 2-4 Customers Under 25 query

Query 3

Create a query called Popular Jobs that counts the number of times each bike repair job is performed. The query should include columns for Job Description and Number of Times Performed. Sort the output so that the most popular jobs are shown at the top of the list. Note the column heading change from the default setting provided by the query generator. Your output should resemble the format shown in Figure 2-5, but the data will be different.

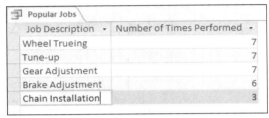

Source: Microsoft product screenshots used with permission from Microsoft Corporation

FIGURE 2-5 Popular Jobs query

Query 4

Create an update query that increases the price of each job by 5 percent. Save the query as Increased Prices. Run the query to test it and view the records in the table to ensure that it works properly.

Report 1

Create a report named Total Owed by Customer that summarizes how much money customers owe for each job performed to fix their bicycles. The report's output should show headings for Last Name, First Name, Street Address, City, Date, Job Description, and Price. You need to create a query first to bring all the fields together. Group the report on the customer's Last Name and total the Price for each group of repairs. Adjust your output so that the Last Name, First Name, and Street Address columns are on the same line and all fields are formatted and visible, as shown in Figure 2-6. Depending on your data, the output should resemble that shown in Figure 2-6. Note that only a portion of the report appears in the figure.

Total Owed by Customer						
Last Name	First Name	Street Address	City	Date	Job Description	Price
Almquist	Antonia	406 Gypsy Hill Road	Longview			
				3/8/2016	Gear Adjustment	$19.53
				3/8/2016	Wheel Trueing	$12.69
				3/8/2016	Tune-up	$31.13
						$63.35
Dawson	James	76 North East Lane	Longview			
				3/8/2016	Tune-up	$31.13
				3/8/2016	Brake Adjustment	$18.47
				3/8/2016	Gear Adjustment	$19.53
						$69.13
Greene	Lauren	67 Main Street	Seattle			
				3/8/2016	Wheel Trueing	$12.69
				3/8/2016	Tune-up	$31.13
				3/8/2016	Brake Adjustment	$18.47
				3/8/2016	Gear Adjustment	$19.53
						$81.82

Source: Microsoft product screenshots used with permission from Microsoft Corporation

FIGURE 2-6 Total Owed by Customer report

Report 2

Create a report called Work Report that summarizes all work performed by the technicians. The report includes columns for Name, Job Description, Date, and Price. You need to create a query for this report first. Group the report by the Name column and include subtotals for the price of each group of repairs, as shown in Figure 2-7. Note that only a portion of the report appears in the figure and that your data will vary.

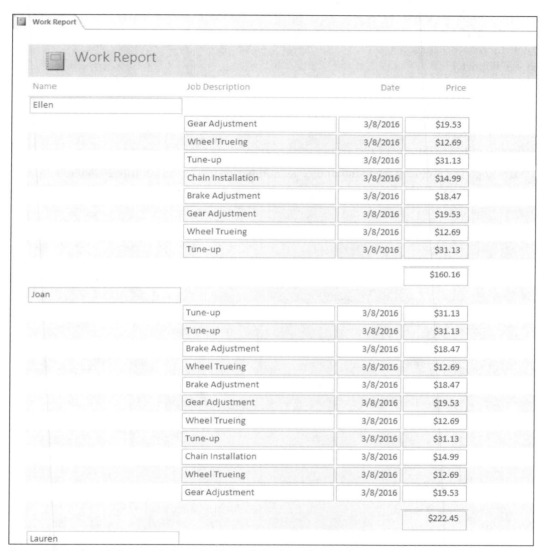

Source: Microsoft product screenshots used with permission from Microsoft Corporation

FIGURE 2-7   Work report

ASSIGNMENT 3: MAKING A PRESENTATION

Create a presentation that you will use to explain the database to Uncle Joe and any clerical staff. Include the design of your database tables and instructions for using the database. Discuss future improvements to the database, such as the ability to update prices for particular jobs and target different customer segments other than university students. Your presentation should take less than 10 minutes, including a brief question-and-answer period.

DELIVERABLES

Assemble the following deliverables for your instructor, either electronically or in printed form:

1. Word-processed design of tables
2. Tables created in Access
3. Form and subform: Visits
4. Query 1: Jobs by Technician Number
5. Query 2: Customers Under 25
6. Query 3: Popular Jobs
7. Query 4: Increased Prices
8. Query for Report 1
9. Report 1: Total Owed by Customer
10. Query for Report 2
11. Report 2: Work Report
12. Presentation materials
13. Any other required printouts or electronic media

Staple all the pages together. Write your name and class number at the top of each page. Make sure that your electronic media are labeled, if required.

THE PARTY RENTALS DATABASE

Designing a Relational Database to Create Tables, Forms, Queries, and Reports

PREVIEW

In this case, you will design a relational database for a company that rents party equipment such as tents and tables. After your design is completed and correct, you will create database tables and populate them with data. You will produce one form with a subform that allows you to record incoming rentals. You will also produce eight queries and two reports. Five queries will address the following questions: Which customers rented a large tent? For how many dates has the small tent been rented? What products are available, and what are their prices? Which are the top-grossing products? Who are the best customers? Another query will update the prices of the tents. Your reports, both based on queries, will summarize all rentals and workers' hours.

PREPARATION

- Before attempting this case, you should have some experience in database design and in using Microsoft Access.
- Complete any part of Tutorial A that your instructor assigns.
- Complete any part of Tutorial B that your instructor assigns, or refer to the tutorial as necessary.
- Refer to Tutorial F as necessary.

BACKGROUND

Your instructor has asked you to help design a database for a company called Party Rentals. She rented several items from the company for a summer backyard party and began talking with the owners, who mentioned that they keep all records on paper and would like to transition to a computer database system. Your instructor introduced you to the owners, and they hired you for a summer internship. They would like your help in designing a prototype database for the business.

The company rents all sorts of party equipment for use in a private home or banquet hall. For example, the company has several large barbecue grills that can accommodate 40 hamburgers at one time. Party Rentals also offers tables, chairs, glassware, plates, tablecloths, tents, and other party items. Each item has a flat rental fee for a 48-hour period. Payment for the rentals must be made in advance—when the reservation is made, an invoice is sent immediately and must be paid before any items are delivered to the home or banquet hall. Your design must therefore include some means of indicating whether a reservation has been paid. All rental costs include delivery, setup, and pickup. Note that deliveries might be made to a different address from that of the customer.

Party Rentals employs several people to run their business. The owner, Darlene, and her husband, Harry, run the office and take care of the paperwork. They have not disclosed the employees' salaries and ask that you not include them in the database design. The couple employs five part-time workers to deliver, set up, and pick up the rental items. These workers are paid on an hourly basis.

In the spring, workers are so busy that they might work several different jobs in one day. For example, they might begin in the morning by setting up a church picnic and then set up for a wedding in a banquet hall later that evening.

To design the database properly, you are shown a sample invoice that displays customer Ellen Monk's rental order (see Figure 3-1). This invoice should help you understand what sorts of information you need to capture in your design. For simplicity's sake, assume there is no sales tax on rentals.

Party Rentals Inc.   **INVOICE**

25 Main Street
Newark DE 19711   (302) 565-5555

Sold to: Invoice number | 536524
Ellen Monk   Invoice date | February 25, 2014
9 Purnell Street   Our order no. | 726278
Newark, DE 19716   Your order no. | 1892727

Delivered to: Sales rep | Joe
Fair Hill Banquet Hall
Elkton MD 21921

Sales tax rate   0.00%

Quantity	Description	Unit price	Amount
1	Extra large BBQ grill	50.00	$50.00
10	5 foot round table	10.00	100.00
60	Folding chairs	5.00	300.00
1	Large tent	100.00	100.00
60	Large plastic beverage glasses	1.00	60.00

	Subtotal	610.00
	Tax	0.00
	Freight	

Direct all inquiries to: Make all checks payable to: Pay this $610.00
Name   Party Rentals Inc.   Amount
(302) 565-5555   Attn: Accounts Receivable
Email: someone@somename.com   25 Main Street
                                Newark, DE 19711

Thank you for your business!

Source: © 2015 Cengage Learning®

FIGURE 3-1   Sample invoice

Once the database is designed and implemented, Darlene and Harry would like to organize various types of information in forms, queries, and reports. For example, they would like an easy way to input reservations in a form. They also need to produce information for planning purposes, which you state they can handle by creating queries. For example, Darlene and Harry want to know the number of dates the small tent has been rented, in case they need to purchase more. They would like to see which products are their top-grossing rentals. Of course, they also want to be able to track their best customers.

Darlene and Harry are also interested in knowing which customers rented particular items. For example, a customer returned a large tent with a rip in it, so Darlene and Harry want to be able to trace the order and charge the customer for the damage. Also, potential customers often call and want to know which items they can rent within a specific range of prices. You explain to the owners that queries can handle both requests easily.

Finally, you suggest two reports that would be useful for the company's decision making. The first one contains information about customers, what items they rented, and the total amount of money they spent. The other report lists all workers and how many hours they have worked during a specific period.

ASSIGNMENT 1: CREATING THE DATABASE DESIGN

In this assignment, you design your database tables using a word-processing program. Pay close attention to the logic and structure of the tables. Do not start developing your Access database in Assignment 2 before getting feedback from your instructor on Assignment 1. Keep in mind that you need to examine the requirements in Assignment 2 to design your fields and tables properly. It is good programming practice to look at the required outputs before beginning your design. When designing the database, observe the following guidelines:

- First, determine the tables you will need by listing the name of each table and the fields it should contain. Avoid data redundancy. Do not create a field if it can be created by a calculated field in a query.
- You will need transaction tables. Think about the business events that occur with each customer's rental. Record rentals for only one specific date. Avoid duplicating data.
- Keep in mind that customers typically rent more than one product for a single order.
- Document your tables using the table feature of your word processor. Your tables should resemble the format shown in Figure 3-2.
- You must mark the appropriate key field(s) by entering an asterisk (*) next to the field name.
- Print the database design if your instructor requires it.

Table Name	
Field Name	Data Type (text, numeric, currency, etc.)
...	...
...	...

Source: © 2015 Cengage Learning®

FIGURE 3-2   Table design

> **NOTE**
>
> Have your design approved before beginning Assignment 2; otherwise, you may need to redo Assignment 2.

ASSIGNMENT 2: CREATING THE DATABASE, QUERIES, AND REPORT

In this assignment, you first create database tables in Access and populate them with data. Next, you create a form, queries, and reports.

Assignment 2A: Creating Tables in Access

In this part of the assignment, you create your tables in Access. Use the following guidelines:

- Create at least nine customer records and add yourself as the tenth customer.
- Create 10 different types of products to rent.
- Create at least 10 rental orders for a single day. Enter data for that day only. Make sure that most customers rent more than one product and that they rent multiple quantities of some products.
- Appropriately limit the size of the text fields; for example, a telephone number does not need the default length of 255 characters.
- Print all tables if your instructor requires it.

Assignment 2B: Creating Forms, Queries, and Reports

You must generate one form with a subform, eight queries, and two reports, as outlined in the Background section of this case.

Form

Create a form and subform based on the table of orders you developed. Save the form as Orders. Your form should resemble that shown in Figure 3-3.

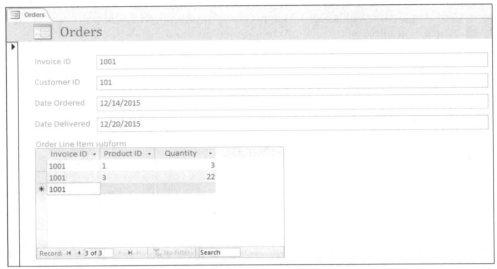

Source: Microsoft product screenshots used with permission from Microsoft Corporation

FIGURE 3-3   Orders form and Order Line Item subform

Query 1

The owners would like to notify all customers who booked a 25-foot tent that it has a rip and needs to be replaced. Create a query that includes columns for the Product Name and the Last Name, First Name, and Email of customers who rented 25-foot tents. Save the query as Customers who Rented Large Tent? Your output should resemble that shown in Figure 3-4, although your data will be different.

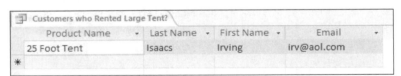

Source: Microsoft product screenshots used with permission from Microsoft Corporation

FIGURE 3-4   Customers who Rented Large Tent? query

Query 2

The owners are curious to know the number of dates the small tent has been rented. Create a query that produces the number. Display columns for the Product Name and Number of Dates Rented as the output to the query. Save the query as Number of Dates Small Tent Rented. Your output should resemble that shown in Figure 3-5, but your data will be different.

Source: Microsoft product screenshots used with permission from Microsoft Corporation

FIGURE 3-5   Number of Dates Small Tent Rented query

Query 3

The owners need to track how much money they are grossing from the rental products. Create a query that lists each product name and the revenue it has generated. The revenue total can be calculated from the price of each rental item and the number of items rented. List the top-grossing product first, as shown in Figure 3-6. Save your query as Top Grossing Products.

Product Name	Revenue Generated
10 Foot Tent	$5,500.00
25 Foot Tent	$3,000.00
5 Foot Round Table	$450.00
White Folding Chair	$365.00
Large Plate	$335.00
8 Foot Rectangular Table	$287.50
Large BBQ Grill	$250.00
Silver Place Setting	$80.00
Dessert Plate	$46.50
Large Glass Tumbler	$31.50

Source: Microsoft product screenshots used with permission from Microsoft Corporation

FIGURE 3-6   Top Grossing Products query

Query 4

The business wants to increase tent rentals, so it is reducing its tent prices by 2 percent. Create an update query called Updated Tent Prices to make the reduction. Run the query to test it, and check the table to ensure that the data is properly updated.

Query 5

Create a query called Products by Price, which is useful when potential customers call and ask for a list of products that fall within a certain price range. The query should prompt for an upper limit of the price range. For example, the output shown in Figure 3-7 displays the names and prices of all items that cost $50 or less. Save the query as Products by Price.

Product Name	Price
Large BBQ Grill	$50.00
5 Foot Round Table	$10.00
White Folding Chair	$5.00
Large Plate	$2.50
Dessert Plate	$1.50
Large Glass Tumbler	$0.50
Silver Place Setting	$2.50
8 Foot Rectangular Table	$12.50

Source: Microsoft product screenshots used with permission from Microsoft Corporation

FIGURE 3-7   Products by Price query

Query 6

Create a query that lists all customers and how much each has spent on rentals. Sort the output from the most money spent to the least. Display columns for Last Name, First Name, Address, City, State, Zip, Telephone, Email, and Total Spent. Note the column heading change from the default setting provided by the query generator. Save the query as Best Customers. Your output should resemble the format shown in Figure 3-8, but the data will be different.

Best Customers

Last Name	First Name	Address	City	State	Zip	Telephone	Email	Total Spent
Downing	Charlie	200 Mac Duff Rd	Austin	TX	78701	(801)912-6564	charlieboy@comcast	$5,205.00
Isaacs	Irving	10010 N. Barrett Dr	Austin	TX	78701	(201)584-1028	irv@aol.com	$3,210.00
DeVivo	Paul	625 Dawson Rd	Austin	TX	78701	(985)453-0943	tanz@zoom.net	$590.00
Bevans	Beverly	24 Raway Rd	Austin	TX	78701	(734)452-0914	bev@comcast.net	$430.00
Smith	John	34 Redback Road	Austin	TX	78701	(512)998-0675	Jsmith@pcp.net	$205.00
Trainer	Lisa	9000 Rt 40, Apt 25-A	Austin	TX	78701	(415)217-9065	readinglisa@aol.com	$196.00
Thompson	Cynthia	31 Spectrum Drive	Austin	TX	78701	(541)443-1123	loohoo@zoom.net	$137.50
Chavez	Mary	9 Elm Arch Road	Austin	TX	78701	(508)443-3212	Jmary@comcast.net	$122.00
O'Hara	Heather	20 Cheswold Blvd	Austin	TX	78701	(307)887-0176	russ101@aol.com	$80.00

Source: Microsoft product screenshots used with permission from Microsoft Corporation

FIGURE 3-8 Best Customers query

Report 1

Create a report called Sales Report. First, create a query that will feed into this report; name the query Sales Report as well. Display columns for the Last Name, First Name, and Address of the customer, and for the Product Name, Quantity, and Total Price of the rental item. Total Price is a calculated field. Bring the query into the report generator and group the output by Last Name, adjusting the output as shown in Figure 3-9. Include subtotals for the total price charged to each customer. All headings should be visible and all data should be formatted correctly. Your output should resemble that shown in Figure 3-9, but your data will be different; only the top portion of the report is shown.

Sales Report

Sales Report

Last Name	First Name	Address	Product Name	Quantity	Total Price
Bevans	Beverly	24 Raway Rd			
			Large BBQ Grill	2	$100.00
			5 Foot Round Table	33	$330.00
					$430.00
Chavez	Mary	9 Elm Arch Road			
			White Folding Chair	11	$55.00
			Dessert Plate	31	$46.50
			Large Glass Tumbler	41	$20.50
					$122.00
DeVivo	Paul	625 Dawson Rd			
			Large Plate	112	$280.00
			White Folding Chair	62	$310.00
					$590.00

Source: Microsoft product screenshots used with permission from Microsoft Corporation

FIGURE 3-9 Sales report

Report 2

Create a report called Workers' Hours. First, create a query that will feed into this report; name the query Workers' Hours as well. Display columns for the last name, first name, and e-mail address of each worker, along with the dates they worked, the clock-in time of each shift, and the total hours worked in each shift (a calculated field). Bring the query into the report generator and group the output by Last Name, adjusting the output as shown in Figure 3-10. Include subtotals for each worker's hours. All headings should be visible and all data should be formatted correctly. Your output should resemble that shown in Figure 3-10, but your data will be different; only the top portion of the report is shown.

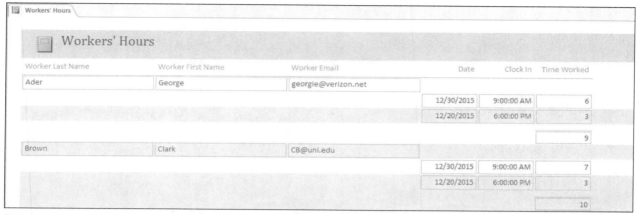

Source: Microsoft product screenshots used with permission from Microsoft Corporation

FIGURE 3-10   Workers' Hours report

ASSIGNMENT 3: MAKING A PRESENTATION

Create a presentation that explains the database to Harry and Darlene. Demonstrate how they can use the database by running the queries and generating reports. Discuss future improvements and additions to the database, and how they might use it to expand their rental business. Your presentation should take less than 10 minutes.

DELIVERABLES

Assemble the following deliverables for your instructor, either electronically or in printed form:

1. Word-processed design of tables
2. Tables created in Access
3. Form and subform: Orders
4. Query 1: Customers who Rented Large Tent?
5. Query 2: Number of Dates Small Tent Rented
6. Query 3: Top Grossing Products
7. Query 4: Updated Tent Prices
8. Query 5: Products by Price
9. Query 6: Best Customers
10. Query 7: Sales Report
11. Query 8: Workers' Hours
12. Report 1: Sales Report
13. Report 2: Workers' Hours
14. Presentation materials
15. Any other required printouts or electronic media

Staple all the pages together. Include your name and class number at the top of each page. Make sure that your electronic media are labeled, if required.

THE BOAT RENTAL DATABASE

Designing a Relational Database to Create Tables, Forms, Queries, and Reports

PREVIEW

In this case, you will design a relational database for a business that arranges boat rentals from private owners. After your design is completed and corrected, you will create database tables and populate them with data. Then you will produce one form with a subform, six queries, and one report. The queries will answer questions such as which renters are not yet qualified to rent a boat, which powerboats in the fleet are a certain length, how much money the boats' owners have made from the rentals, how many days the boats have been rented, and how many times they have been taken out. Another query will update rental prices. You will produce one report based on a query that displays the rental summary.

PREPARATION

- Before attempting this case, you should have some experience in database design and in using Microsoft Access.
- Complete any part of Tutorial A that your instructor assigns.
- Complete any part of Tutorial B that your instructor assigns, or refer to the tutorial as necessary.
- Refer to Tutorial F as necessary.

BACKGROUND

You spent your summers as a child boating on Chesapeake Bay with your family and friends, and as a young adult you worked at a marina. You know from experience that people often have large investments in their boats but don't use them frequently. The increased popularity of smartphones has enabled the growth of services such as shared taxis, and you think that boat sharing could be a viable business as well. Working with your friend Amy, you have decided to create a sharing service to connect boat owners with people who want to rent boats for a short time. After spending the summer working on the business idea and linking a few boat owners with potential renters, you feel that the company could grow and thrive. Before you can get funding for the business, however, you must prove to the investors that it will be successful. To prepare for your first meeting with potential investors, you decide to create a prototype database using Microsoft Access to show them how the business will work.

You meet with Amy to figure out how the business works and how to support it with the database. The two of you envision boat owners registering with your business and providing information such as the type of boat they want to rent out (powerboat or sailboat), its brand, and its length in feet. You also think that the owner should set the price of the rental and that you should advertise rentals at a daily rate. For example, a large, fancy powerboat might cost $450 per day. To make the business worthwhile, you and Amy think the owner should give the business a cut of the rental income, but you are not certain what you should charge the owners. At this point, you will not program such payments into the prototype database.

Potential renters can register with the business for free and then browse the selection of boats that are available for rent. You and Amy are hoping that many people across the country will register as potential renters. Each renter's contact information will be recorded, along with information about the boat they rent and the starting and ending dates of the rental.

After you set up the database, you would like to include several useful features that help the investors understand how the business will work. First, you think that a form and subform would be an efficient way of recording boat rentals.

You and Amy then brainstorm to determine what questions the database should answer. These questions will become the queries you set up for demonstration purposes. First, safety is the utmost priority, so you feel no one should rent to a potential boater who is not qualified. One query will list all potential renters who have not yet been fully qualified for boating.

Amy convinces you that users of the database will want to search for boats in different ways. For example, powerboat renters might be interested in boats of a particular length. You know that a query is an excellent way to find such information. Amy also thinks you need to be able to change rental prices for different types of boats. You decide to include an update query that changes the rental price of all sailboats so investors comprehend the power of the database system.

The investors need to know how much money boat owners can make from the rentals. You decide to create a query that calculates the amount of money earned by each boat owner. As additional support for the business, you want to include queries that list the number of days each boat is rented and the number of times each boat is rented.

Finally, you will complete the prototype database by developing a nicely formatted report that summarizes all boat rentals.

ASSIGNMENT 1: CREATING THE DATABASE DESIGN

In this assignment, you design your database tables using a word-processing program. Pay close attention to the logic and structure of the tables. Do not start developing your Access database in Assignment 2 before getting feedback from your instructor on Assignment 1. Keep in mind that you need to examine the requirements in Assignment 2 to design your fields and tables properly. It is good programming practice to look at the required outputs before beginning your design. When designing the database, observe the following guidelines:

- First, determine the tables you will need by listing the name of each table and the fields it should contain. Avoid data redundancy. Do not create a field if it can be created by a calculated field in a query.
- You will need a transaction table. Think about the business events that occur with boat rentals. Avoid duplicating data.
- Document your tables using the table feature of your word processor. Your tables should resemble the format shown in Figure 4-1.
- You must mark the appropriate key field(s) by entering an asterisk (*) next to the field name. Keep in mind that some tables might need a compound primary key to uniquely identify a record within a table.
- Print the database design, if required.

Table Name	
Field Name	Data Type (text, numeric, currency, etc.)
...	...
...	...

Source: © 2015 Cengage Learning®

FIGURE 4-1   Table design

NOTE

Have your design approved before beginning Assignment 2; otherwise, you may need to redo Assignment 2.

ASSIGNMENT 2: CREATING THE DATABASE, QUERIES, AND REPORTS

In this assignment, you first create database tables in Access and populate them with data. Next, you create a form, six queries, and one report.

Assignment 2A: Creating Tables in Access

In this part of the assignment, you create your tables in Access. Use the following guidelines:

- Enter data for at least 10 boat owners and 10 renters. Use your name as one of the owners or renters. Enter address and e-mail information for each party involved. Make sure one of the renters is not qualified to drive a boat.
- Create records for more than 10 boats so that some of the owners have more than one boat to rent.
- Appropriately limit the size of the text fields; for example, a phone number does not need the default length of 255 characters.
- Print all tables if your instructor requires it.

Assignment 2B: Creating Forms, Queries, and Reports

You will generate one form with a subform, six queries, and one report, as outlined in the Background section of this case.

Form

Create a form and subform based on your Boats table and Rentals table (or whatever you named the tables). Save the form as Boats. Your form should resemble that in Figure 4-2. Note that only a portion of the Rentals subform appears; the rest of the records become visible when you scroll.

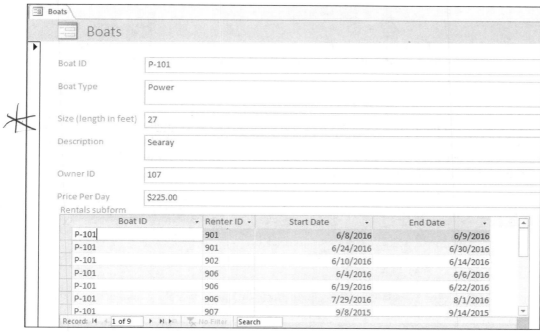

Source: Microsoft product screenshots used with permission from Microsoft Corporation

FIGURE 4-2 Boats form and subform

Query 1

Create a select query called Renters not Qualified that lists the renter's ID number, name, address, telephone number, and e-mail address. Your output should resemble that in Figure 4-3, although your data will be different.

Renters not Qualified								
Renter ID	Renter Last Name	Renter First Name	Renter Street Address	Renter City	Renter State	Renter Zip	Renter Telephone	Renter Email
904	Denron	Susan	410 N. Church Street	Tampa	FL	33160	403-887-2736	susand@verizon.net

Source: Microsoft product screenshots used with permission from Microsoft Corporation

FIGURE 4-3   Renters not Qualified query

Query 2

Create a parameter query called Power Boats by Length. The query prompts the user to enter a lower limit for the size of a powerboat and then displays headings for Boat Type, Size, and Description. For example, if you enter 27 when prompted, the output should resemble that in Figure 4-4, although your data will be different.

Boat Type	Size (length in feet)	Description
Power	27	Searay
Power	28	Active Thunder Savage
Power	30	Betram Sedan
*	0	

Source: Microsoft product screenshots used with permission from Microsoft Corporation

FIGURE 4-4   Power Boats by Length query

Query 3

Create an update query that lowers the price of sailboat rentals by 2 percent. Save the query as Lower Sailboat Prices. Run the query to make sure it works properly and then confirm the results by viewing the data in the Boats form.

Query 4

Create a query that calculates the amount of money each boat owner earns from <u>rentals</u>. Display the names and addresses of the boat owners and calculate the amount of money they have earned from rentals. Sort the query results to list the greatest earnings first. Save the query as Money Earned by Owners. Your output should resemble the format shown in Figure 4-5, but the data will be different.

Money Earned by Owners						
Last Name	First Name	Address	City	State	Zip	Money Earned
Seals	Lauren	163 Darling Rd	Kennet Square	PA	19348	$20,550.00
Mattern	Luis	205 Hanover Pl	Avondale	PA	19311	$12,500.00
Webster	Sally	15 Anglin Dr	Wilmington	DE	19808	$9,800.00
Lee	Gregory	139 Boyer Way	Avondale	PA	19311	$8,250.00
Meartz	Maria	511 Sparrow Ct	Elkton	MD	21921	$8,200.00
Burham	Luke	212 Wedgewood Rd	Kennet Square	PA	19348	$7,525.00
Seever	Patti	19 Danvers Circle	Kennet Square	PA	19348	$7,425.00
Ward	Harry	11 Chapel Rd	Wilmington	DE	19808	$6,460.00
Poplawski	Meredith	281 Beverly Rd	Elkton	MD	21119	$5,250.00
Doorey	Leon	9 Sandelwood Dr	Avondale	PA	19311	$3,125.00

Source: Microsoft product screenshots used with permission from Microsoft Corporation

FIGURE 4-5   Money Earned by Owners query

(Price per day × days Rented)

3528
Alexander

1372
431.2
1729.8

Query 5

Create a query called Number of Days Boats Rented that displays columns for the Boat Type, Size, and Description and calculates the Number of Days Rented. Sort the query results so that the boat rented most often appears at the top of the list. Your output should resemble the format shown in Figure 4-6, but the data will be different.

Number of Days Boats Rented			
Boat Type ▾	Size (length in feet) ▾	Description ▾	Number of Days Rented ▾
Sail	27	J Boat	55
Sail	29	Hunter	50
Power	21	Baja Outlaw	43
Sail	32	Catalina	41
Sail	35	Island Packet	34
Power	28	Active Thunder	34
Power	27	Searay	33
Power	30	Betram Sedan	28
Power	18	Apache Scout	25
Sail	32	Beneteau	20
Sail	45	Tartan	15

Source: Microsoft product screenshots used with permission from Microsoft Corporation

FIGURE 4-6   Number of Days Boats Rented query

Query 6

Create a query called Number of Times Boats Rented that displays columns for the Boat ID, Boat Type, and Size and calculates the Number of Times Rented. Sort the query results so that the boat rented most often appears at the top of the list. Your output should resemble the format shown in Figure 4-7, but the data will be different.

Number of Times Boats Rented			
Boat ID ▾	Boat Type ▾	Size (length in feet) ▾	Number of Times Rented ▾
S-101	Sail	27	15
S-105	Sail	29	12
P-105	Power	21	11
S-102	Sail	32	10
P-102	Power	28	10
P-101	Power	27	9
S-106	Sail	32	8
S-104	Sail	35	7
P-103	Power	30	7
S-103	Sail	45	5
P-104	Power	18	5

Source: Microsoft product screenshots used with permission from Microsoft Corporation

FIGURE 4-7   Number of Times Boats Rented query

Report

Create a report called Rental Summary by Renter that summarizes all boat rentals. First, you need to create a query to amass the required data. The report should include headings for Renter, Boat Type, Description, Start Date of the rental, and Total Price, which is a calculated field. Group the report by the renter's name. Include subtotals that display the total amount spent by each renter. Make sure all headings are visible and that the data is formatted correctly, as shown. Depending on your data, your output should resemble that in Figure 4-8.

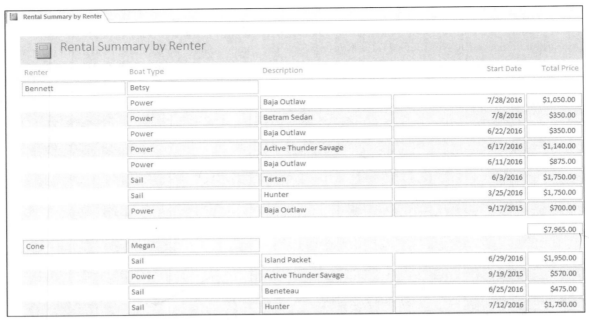

Source: Microsoft product screenshots used with permission from Microsoft Corporation

FIGURE 4-8   Rental Summary by Renter report

ASSIGNMENT 3: MAKING A PRESENTATION

Create a presentation that explains the database to the investors and demonstrates how it is used. Discuss future improvements to the database, such as moving the system online. Your presentation should take less than 10 minutes, including a question-and-answer period.

DELIVERABLES

Assemble the following deliverables for your instructor, either electronically or in printed form:

1.   Word-processed design of tables
2.   Tables created in Access
3.   Form and subform: Boats
4.   Query 1: Renters not Qualified
5.   Query 2: Power Boats by Length
6.   Query 3: Lower Sailboat Prices
7.   Query 4: Money Earned by Owners
8.   Query 5: Number of Days Boats Rented
9.   Query 6: Number of Times Boats Rented
10. Query 7: for report
11. Report: Rental Summary by Renter
12. Presentation materials
13. Any other required printouts or electronic media

Staple all the pages together. Include your name and class number at the top of the page. Make sure that your electronic media are labeled, if required.

THREE WHEELED DELIVERIES DATABASE

Designing a Relational Database to Create Tables, Queries, and Reports

PREVIEW

In this case, you will design a relational database for a company that delivers produce, baked goods, coffee, and office supplies throughout New York City using electrically powered tricycles. After your tables are designed and created, you will populate the database and create a form with a subform, five queries, and two reports. The form and subform will allow for easy submission of hours worked by employees. The queries will address the following questions: What deliveries are made to a specific company on a specific day? How many times has each tricycle been used? What is the location of a specific package? How many package deliveries were made on a specific date? How many packages have been shipped by each business? The reports will summarize the hours worked by employees and the number of packages delivered for each client company.

PREPARATION

- Before attempting this case, you should have some experience in database design and in using Microsoft Access.
- Complete any part of Tutorial A that your instructor assigns.
- Complete any part of Tutorial B that your instructor assigns, or refer to the tutorial as necessary.
- Refer to Tutorial F as necessary.

BACKGROUND

Many cities are investing in bike lanes for safe bicycle travel, and companies are interested in reducing pollution and greenhouse gases in cities. The two ideas have coalesced in Three Wheeled Deliveries (TWD), a new company in New York City that uses electrically powered full-sized tricycles to make local deliveries of produce, baked goods, coffee, and office supplies. Businesses pay a little more for the delivery service in exchange for contributing to the health and well-being of New York City residents. In addition, they are setting a good role model for other businesses.

Your Uncle Nate is the founder of TWD. He meets you over the Thanksgiving break and talks to you about his fledgling business. You tell him about your information systems class and how you have learned to design databases and implement them using Microsoft Access. Later that weekend, Uncle Nate e-mails you and offers you a job over the upcoming winter holiday to help TWD design and implement a prototype database. You agree to take on the challenge and the opportunity to make some needed cash as well.

On your first day of work, Nate explains how the business operates. So far, TWD has four delivery clients in New York City. Those businesses pack their goods in boxes of different sizes to fit neatly into the trailer that is pulled behind the tricycle. Each box of goods is weighed, the weight is recorded, and the package is given a unique tracking number. The names of the sending business and the package recipient are recorded as well. (Nate is sensitive about divulging too much data, so he calls the recipients Company A, Company B, and so on.) A large number of boxes can fit into one trailer, so each delivery run can include many packages.

TWD has hired 10 people to ride the trikes around the city and deliver packages. These riders work part time and only one rider works per day, averaging about eight hours of delivery time each shift. Riders must clock in and out each day because their shifts do not conform to rigid starting and ending times. One trike load is delivered each day, but Nate hopes that TWD will soon grow to deliver four daily trike loads. He owns three trikes that are brightly painted. When the delivery person takes a trike, Nate records whether it is red, blue, or green. The delivery number is also included in Nate's records, along with the rider's name and the date.

Once the prototype database is designed and the tables are populated with data, Nate has a few additional requests. First, he would like to provide workers with an easy way to enter their clock-in and clock-out times directly into the database. You explain that a form and subform will easily handle the chore.

Nate also has several frequently asked questions that he wants the database to answer. For example, TWD's clients need to know which deliveries went to certain destinations on a particular day. To help monitor his business, Nate wants to know how often each trike is used so he can ensure proper maintenance. Also, businesses sometimes call and need to know whether a package has been delivered. If it has, they need to know the delivery date.

For analysis purposes, Nate wants to monitor certain data at TWD. Specifically, he would like to know the total number of packages shipped from each business and the number of deliveries made each day. You are confident that all his questions can be handled by Access queries.

Nate needs to keep close tabs on the business and would like to have summary reports to help solicit investors. Specifically, he would like to have a report that summarizes the number of deliveries from each of his client businesses and indicates which customers receive the deliveries. Nate would also like to have a neatly formatted report that lists each employee and how many hours they worked on specific days. Again, you are confident that the database can easily handle his requests for reports.

ASSIGNMENT 1: CREATING THE DATABASE DESIGN

In this assignment, you design your database tables using a word-processing program. Pay close attention to the logic and structure of the tables. Do not start developing your Access database in Assignment 2 before getting feedback from your instructor on Assignment 1. Keep in mind that you need to examine the requirements in Assignment 2 to design your fields and tables properly. It is good programming practice to look at the required outputs before beginning your design. When designing the database, observe the following guidelines:

- First, determine the tables you will need by listing the name of each table and the fields it should contain. Avoid data redundancy. Do not create a field if it can be created by a calculated field in a query.
- You will need a few transaction tables to record all deliveries and the packages that make up each delivery. Transactions also occur with the employees' working hours.
- Document your tables using the table feature of your word processor. Your tables should resemble the format shown in Figure 5-1.
- You must mark the appropriate key field(s) by entering an asterisk (*) next to the field name. Keep in mind that some tables might need a compound primary key to uniquely identify a record within a table.
- Print the database design.

Table Name	
Field Name	Data Type (text, numeric, currency, etc.)
…	…
…	…

Source: © 2015 Cengage Learning®

FIGURE 5-1   Table design

NOTE

Have your design approved before beginning Assignment 2; otherwise, you may need to redo Assignment 2.

ASSIGNMENT 2: CREATING THE DATABASE, FORM, QUERIES, AND REPORTS

In this assignment, you first create database tables in Access and populate them with data. Next, you create a form and subform, five queries, and two reports.

Assignment 2A: Creating Tables in Access

In this part of the assignment, you create your tables in Access. Use the following guidelines:

- Create records for four businesses that ship with TWD.
- Create records for 10 different workers with fictitious names, addresses, telephone numbers, and e-mail addresses, as outlined in the Background section of this case.
- Create records for at least 100 packages with unique tracking numbers and various weights and sizes. Limit the box sizes to small, medium, and large. Consider using Microsoft Excel to generate random data and alleviate user typing. You can use the =randbetween function in Excel to create random data.
- Create deliveries for 13 days and scatter the transactions for all packages among those 13 days. Have each employee work at least one day, and have some employees work two days. Have each worker clock in and out for shifts that are close to eight hours per day.
- Create records for three trikes—one red, one blue, and one green, orange, yellow, violet
- Appropriately limit the size of the text fields; for example, a customer ID number does not need the default length of 255 characters.
- Print all tables if your instructor requires it. Upload on BB

Assignment 2B: Creating a Form, Queries, and Reports

You will create a form and subform, five queries, and two reports, as outlined in the Background section of this case.

Form

Create a form and subform based on your Delivery Personnel table and Delivery Person Hours Worked table (or whatever you called those tables.) Save the form as Delivery Personnel.

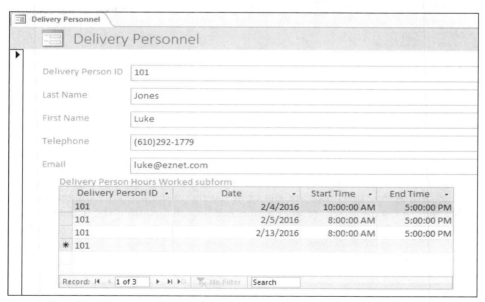

Source: Microsoft product screenshots used with permission from Microsoft Corporation

FIGURE 5-2   Delivery Personnel form and subform

Query 1

Create a query called "Deliveries to a specific company on a specific date" that shows all deliveries to a requested company on a requested date. Display columns for Company, Date Delivered, Delivery Person, and Trike. Choose one company and one date to filter your results. Your data will vary, but the output should resemble that in Figure 5-3.

too many entries?

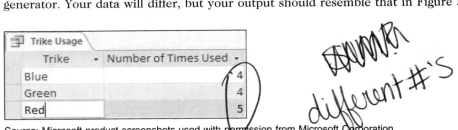

Source: Microsoft product screenshots used with permission from Microsoft Corporation

FIGURE 5-3   Deliveries to a specific company on a specific date

Query 2 ✓

Create a query called Trike Usage that displays columns for each trike and the number of times it has been used, which is a calculation. Note the column heading change from the default setting provided by the query generator. Your data will differ, but your output should resemble that in Figure 5-4.

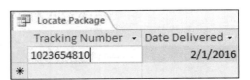

answer different #'s

Source: Microsoft product screenshots used with permission from Microsoft Corporation

FIGURE 5-4   Trike Usage query

Query 3 ✓

Create a query called Locate Package that prompts the user to enter a tracking number and then displays that number with the date the package was delivered. Your data will differ, but your output should resemble that in Figure 5-5, which shows tracking number 1023654810 as the input to the query prompt.

Tracking Number	Date Delivered
1023654810	2/1/2016

Source: Microsoft product screenshots used with permission from Microsoft Corporation

FIGURE 5-5   Locate Package query

Query 4 ✓

Create a query called Delivery Count that determines how many packages are delivered each day. Sort the output so that the days with the most package deliveries are shown at the top of the list. Your data will differ, but your output should resemble that in Figure 5-6. Note the column heading change from the default setting provided by the query generator.

way different #'s (handwritten)

Source: Microsoft product screenshots used with permission from Microsoft Corporation

FIGURE 5-6   Delivery Count query

Query 5   ✓ *but double check* (handwritten)

Create a query called Packages Shipped by Business. Display headings for Business Name and Number of Packages Shipped, which is a calculation. The last field is a counted field that is sorted to show which businesses shipped the most packages. Note the column heading change from the default setting provided by the query generator. Your data will differ, but your output should resemble that in Figure 5-7.

Source: Microsoft product screenshots used with permission from Microsoft Corporation

FIGURE 5-7   Packages Shipped by Business query

Report 1

upage 89 (handwritten)

Create a report called Employee Hours Worked. First create a query that includes columns for each employee's Last Name, First Name, Email, and Date of each shift, plus a calculation of the number of hours worked each day. Save the query as Employee Hours Worked. Bring the query into a report and group it by the Last Name column. In the report generator, adjust the Last Name, First Name, and Email columns to be on the same line. Make sure that all field headings are visible and all data is formatted correctly. Your data will differ, but your output should resemble that in Figure 5-8. Note that only a portion of the data appears in the figure.

Format field | to show time (handwritten)

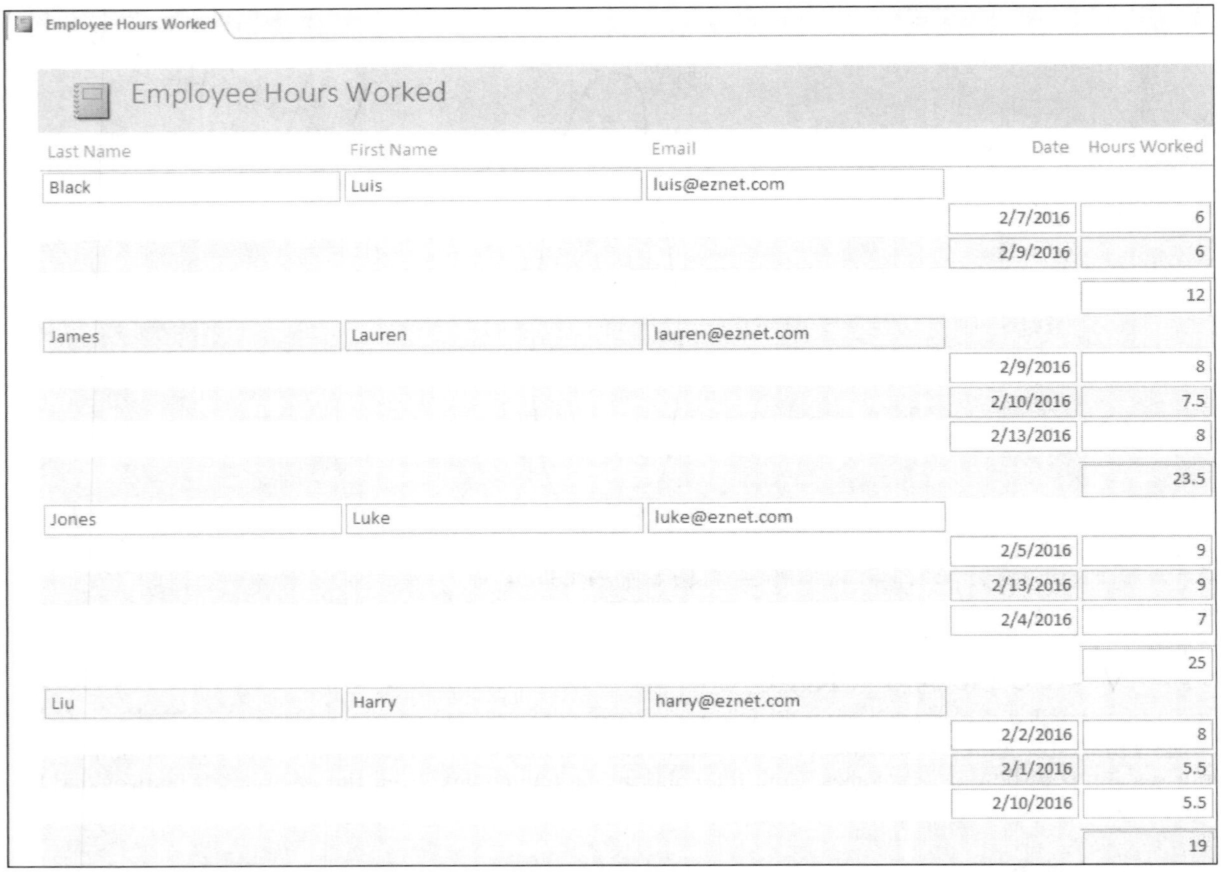

Source: Microsoft product screenshots used with permission from Microsoft Corporation

FIGURE 5-8   Employee Hours Worked report

Report 2

Create a report called Number of Deliveries. First create a query that displays columns for the Business Name (the shipper) and Company (the recipient) and calculates the total number of packages shipped. Save the query as Number of Deliveries, and then bring the query into the report generator and group it by Business Name. Include subtotals that display the number of deliveries for each business. Your data will differ, but your output should resemble that in Figure 5-9.

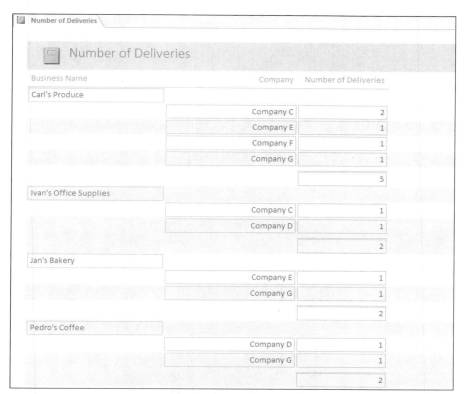

Source: Microsoft product screenshots used with permission from Microsoft Corporation

FIGURE 5-9   Number of Deliveries report

ASSIGNMENT 3: MAKING A PRESENTATION

Create a presentation that explains all the useful functions of your database. Consider discussing how your system could be expanded to record the delivery paths of the trikes and their movements through a GPS. Think of other expansions that would be helpful to Nate. Your presentation should take less than 15 minutes, including a brief question-and-answer period.

DELIVERABLES

Assemble the following deliverables for your instructor, either electronically or in printed form:

1. Word-processed design of tables
2. Tables created in Access
3. Form: Delivery Personnel
4. Query 1: Deliveries to a specific company on a specific date
5. Query 2: Trike Usage
6. Query 3: Locate Package
7. Query 4: Delivery Count
8. Query 5: Packages Shipped by Business
9. Report 1: Employee Hours Worked (includes a query)
10. Report 2: Number of Deliveries (includes a query)

Staple all the pages together. Include your name and class number at the top of the page. Make sure that your electronic media are labeled, if required.

PART **2**

DECISION SUPPORT CASES
USING EXCEL SCENARIO MANAGER

TUTORIAL **C**

BUILDING A DECISION SUPPORT SYSTEM IN EXCEL

Decision Support Systems (DSS) are computer programs used to help managers solve complex business problems. DSS programs are commonly found in large, integrated packages called enterprise resource planning software that provide information services to an organization. Software packages such as SAP™, Microsoft Dynamics™, and PeopleSoft™ offer sophisticated DSS capabilities. However, many business problems can be modeled for solutions using less complex tools, such as Visual Basic, Microsoft Access, and Microsoft Excel.

A DSS program is actually a model representing a quantitative business problem. The problem can range from finding a desired product mix to sales forecasts to risk analysis, but almost all of the problems examine *financial outcomes*. The model itself contains the data and the algorithms (mathematical processes) needed to solve the problem.

In a DSS program, the user manually inputs data or the program accesses data from a file in the system. The program runs the data through its algorithms and displays output formatted as information; the manager uses this data to decide what action to take to solve the problem. Some sophisticated DSS programs display multiple solutions and recommend one based on predefined parameters.

Managers often find the Excel spreadsheet program particularly useful for their DSS needs. Excel contains hundreds of built-in arithmetic, statistical, logical, and financial functions. It can import data in numerous formats from large database programs, and it can be set up to display well-organized, visually appealing tables and graphs from the output.

This tutorial is organized into four sections:

1. **Spreadsheet and DSS Basics**—This section lets you "get your feet wet" by creating a DSS program in Excel. The program is a cash flow model for a small business looking to expand. You will get an introduction to spreadsheet design, building a DSS, and using financial functions.

2. **Scenario Manager**—Here you will learn how to use the Excel Scenario Manager. A DSS typically gives you one set of answers based on one set of inputs—the real value of the tool lies in its ability to play "what if" and take a comparative look at all the solutions based on all combinations of the inputs. Rather than inputting and running the DSS several times manually, you can use Scenario Manager to run and display the outputs from all possible combinations of the inputs. The output is summarized on a separate worksheet in the Excel workbook.

3. **Practice Using Scenario Manager**—Next, you will be given a new problem to model as a DSS, using Scenario Manager to display your solutions.

4. **Review of Excel Basics**—This section reviews additional information that will help you complete the spreadsheet cases that follow this tutorial. You will learn some basic operations, logical functions, and cash flow calculations.

SPREADSHEET AND DSS BASICS

You are the owner of a thrift shop that resells clothing and housewares in a university town. Many of your customers are college students. Your business is unusual in that sales actually increase during an economic recession. Your cost of obtaining used items basically follows the consumer price index. It is the end of 2015, and business has been very good due to the continuing recession. You are thinking of expanding your business to an adjacent storefront that is for sale, but you will have to apply for a business loan to finance the purchase. The bank requires a projection of your profit and cash flows for the next two years before it will loan you the money to expand, so you have to determine your net income (profit) and cash flows for 2016 and 2017. You decide that your forecast should be based on four factors: your 2015 sales dollars, your cost of goods sold per sales dollar, your estimates of the underlying economy, and the business loan payment amount and interest rate.

Because you will present this model to your prospective lenders, you decide to use an Income and Cash Flow Statements framework. You will input values for two possible states of the economy for 2016 and 2017: R for a continuing recession and B for a "boom" (recovery). Your sales in the recession were growing at 20% per year. If the recession continues and you expand the business, you expect sales to continue growing at 30% per year. However, if the economy recovers, some of your customers will switch to buying "new," so you expect sales growth for your thrift shop to be 15% above the previous year (only 5% growth plus 10% for the business expansion). If you do not expand, your recession or boom growth percentages will only be 20% and 5%, respectively. To determine the cost of goods sold for purchasing your merchandise, which is currently 70% of your sales, you will input values for two possible consumer price outlooks: H for high inflation (1.06 multiplied by the average cost of goods sold) and L for low inflation (1.02 multiplied by the cost of goods sold).

You currently own half the storefront and will need to borrow $100,000 to buy and renovate the other half. The bank has indicated that, depending on your forecast, it may be willing to loan you the money for your expansion at 5% interest during the current recession with a 10-year repayment compounded annually ("R"). However, if the prime rate drops at the start of 2016 because of an economic turnaround ("B"), the bank can drop your interest rate to 4% with the same repayment terms.

As an entrepreneur, an item of immediate interest is your cash flow position with the additional burden of a loan payment. After all, one of your main objectives is to make a profit (Net Income After Taxes). You can use the DSS model to determine if it is more profitable *not* to expand the business.

Organization of the DSS Model

A well-organized spreadsheet will make the design of your DSS model easier. Your spreadsheet should have the following sections:

- Constants
- Inputs
- Summary of Key Results
- Calculations (with separate calculations for Expansion vs. No Expansion)
- Income and Cash Flow Statements (with separate statements for Expansion vs. No Expansion)

You can also download the spreadsheet skeleton if you prefer; it will save you time. To access this skeleton file, select Tutorial C from your data files, and then select **Thrift Shop Expansion Skeleton. xlsx**.

Figures C-1 and C-2 illustrate the spreadsheet setup for the DSS model you want to build.

	Tutorial Exercise--Collegetown Thrift Shop			
		2015	**2016**	**2017**
Constants				
	Tax Rate	NA	33%	35%
	Loan Amount for Store Expansion	NA	$100,000	NA
Inputs		**2015**	**2016**	**2017**
	Economic Outlook (R=Recession, B=Boom)	NA		NA
	Inflation Outlook (H=High, L=Low)	NA		NA
Summary of Key Results		**2015**	**2016**	**2017**
	Net Income after Taxes (Expansion)	NA		
	End-of-year Cash on Hand (Expansion)	NA		
	Net Income after Taxes (No Expansion)	NA		
	End-of-year Cash on Hand (No Expansion)	NA		
Calculations (Expansion)		**2015**	**2016**	**2017**
	Total Sales Dollars	$350,000		
	Cost of Goods Sold	$245,000		
	Cost of Goods Sold (as a percent of Sales)	70%		
	Interest Rate for Business Loan		NA	NA

Source: Microsoft product screenshots used with permission from Microsoft Corporation

FIGURE C-1 Tutorial skeleton 1

	A	B	C	D
23	**Calculations (No Expansion)**	**2015**	**2016**	**2017**
24	Total Sales Dollars	$350,000		
25	Cost of Goods Sold	$245,000		
26	Cost of Goods Sold (as a percent of Sales)	70%		
27				
28	**Income and Cash Flow Statements (Expansion)**	**2015**	**2016**	**2017**
29	Beginning-of-year Cash on Hand	NA		
30	Sales (Revenue)	NA		
31	Cost of Goods Sold	NA		
32	*Business Loan Payment*	NA		
33	Income before Taxes	NA		
34	Income Tax Expense	NA		
35	Net Income after Taxes	NA		
36	End-of-year Cash on Hand	$15,000		
37				
38	**Income and Cash Flow Statements (No Expansion)**	**2015**	**2016**	**2017**
39	Beginning-of-year Cash on Hand	NA		
40	Sales (Revenue)	NA		
41	Cost of Goods Sold	NA		
42	Income before Taxes	NA		
43	Income Tax Expense	NA		
44	Net Income after Taxes	NA		
45	End-of-year Cash on Hand	$15,000		

Source: Microsoft product screenshots used with permission from Microsoft Corporation

FIGURE C-2   Tutorial skeleton 2

Each spreadsheet section is discussed in detail next.

The Constants Section

This section holds values that are needed for the spreadsheet calculations. These values are usually given to you, and generally do not change for the exercise. However, you can change these values later if necessary; for example, you might need to borrow more or less money for your business expansion (cell C5). For this tutorial, the constants are the Tax Rate and the Loan Amount.

The Inputs Section

The Inputs section in Figure C-1 provides a place to designate the two possible economic outlooks and the two possible inflation outlooks. If you wanted to make these outlooks change by business year, you could leave blanks under both business years. However, as you will see later when you use Scenario Manager, this approach would greatly increase the complexity of interpreting the results. For simplicity's sake, assume that the same outlooks will apply to both 2016 and 2017.

The Summary of Key Results Section

This section summarizes the Year 2 and 3 Net Income after Taxes (profit) and the End-of-year Cash on Hand both for expanding the business and for not expanding. These cells are copied from the Income and Cash Flow Statements section at the bottom of the sheet. Summary sections are frequently placed near the top of a spreadsheet to allow managers to see a quick "bottom line" summary without having to scroll down the spreadsheet to see the final result. Summary sections can also make it easier to select cells for charting.

The Calculations Sections (Expansion and No Expansion)

The following areas are used to compute the following necessary results:

- The Total Sales Dollars, which is a function of the Year 2015 value and the Economic Outlook input
- The Cost of Goods Sold, which is the Total Sales Dollars multiplied by the Cost of Goods Sold (as a percent of Sales)
- The Cost of Goods Sold (as a percent of Sales), which is a function of the Year 2015 value and the Inflation Outlook input

- In addition, the Calculations section for the expansion includes the interest rate, which is also a function of the Economic Outlook input. This interest rate will be used to determine the Business Loan Payment in the Income and Cash Flow Statements section.

You could make these formulas part of the Income and Cash Flow Statements section. However, it makes more sense to use the approach shown here because it makes the formulas in the Income and Cash Flow Statements less complicated. In addition, when you create other DSS models that include unit costing and pricing calculations, you can enter the formulas in this section to facilitate managerial accounting cost analysis.

The Income and Cash Flow Statements Sections (Expansion and No Expansion)

These sections are the financial or accounting "body" of the spreadsheet. They contain the following values:

- Beginning-of-year Cash on Hand, which equals the *prior* year's End-of-year Cash on Hand.
- Sales (Revenue), which in this tutorial is simply the results of the Total Sales Dollars copied from the Calculations section.
- Cost of Goods Sold, which also is copied from the Calculations section.
- Business Loan Payment, which is calculated using the PMT (Payment) function and the inputs for loan amount and interest rate from the Constants and Calculations sections. Note that only the Income and Cash Flow Statement for Expansion includes a value for Business Loan Payment. If you do not expand, you do not need to borrow the money.
- Income before Taxes, which is Sales minus the Cost of Goods Sold; for the expansion scenarios, you also subtract the Business Loan Payment.
- Income Tax Expense, which is zero when there is no income or the income is negative; otherwise, this value is the Income before Taxes multiplied by the Tax Rate from the Constants section.
- Net Income after Taxes, which is Income before Taxes minus Income Tax Expense.
- End-of-year Cash on Hand, which is Beginning-of-year Cash on Hand plus Net Income after Taxes.

Note that this Income and Cash Flow Statement is greatly simplified. It does not address the issues of changes in Inventories, Accounts Payable, and Accounts Receivable, nor any period expenses such as Selling and General Administrative expenses, utilities, salaries, real estate taxes, insurance, or depreciation. Also note that Business Loan Repayment is not usually considered an expense item. Please ignore this diversion from generally accepted accounting principles in the interest of illustrating cash flow.

Construction of the Spreadsheet Model

Next, you will work through three steps to build the spreadsheet model:

1. Make a skeleton or "shell" of the spreadsheet. Save it with a name you can easily recognize, such as TUTC.xlsx or Tutorial C *YourName*.xlsx. When submitting electronic work to an instructor or supervisor, include your last name and first initial in the filename.
2. Fill in the "easy" cell formulas.
3. Then enter the "hard" spreadsheet formulas.

Again, you can use the spreadsheet skeleton if you prefer; it will save you time. Select Tutorial C from your data files, and then select **Thrift Shop Expansion Skeleton.xlsx**. Rename the file if your instructor requires it.

Making a Skeleton or "Shell"

The first step is to set up the skeleton worksheet. The skeleton should have headings, text labels, and constants. Do not enter any formulas yet.

Before you start entering data, you should first try to visualize a sensible structure for your worksheet. In Figures C-1 and C-2, the seven sections are arranged vertically down the page; the item descriptions are in the first column (A), and the time periods (years) are in the next three columns (B, C, and D). This is a widely accepted business practice, and is commonly called a "horizontal analysis." It is used to visually compare financial data side by side through successive time periods.

Because your key results depend on the Income and Cash Flow Statements, you usually set up that section first, and then work upward to the top of the sheet. In other words, you set up the Income and Cash Flow Statements section, then the Calculations section, and then the Summary of Key Results, Inputs, and Constants sections. Some might argue that the Income and Cash Flow Statements should be at the top of the sheet, but when you want to change values in the Constants or Inputs section or examine the Summary of Key Results, it does not make sense to have to scroll to the bottom of the worksheet. When you run the model, you do not enter anything in the Income and Cash Flow Statements—they are all calculations. So, it makes sense to put them last.

Here are some other general guidelines for designing effective DSS spreadsheets:

- Decide which items belong in the Calculations section. A good rule of thumb is that if your items have formulas but do not belong in the Income and Cash Flow Statements, put them in the Calculations section. Good examples are intermediate calculations such as unit volumes, costs and prices, markups, or changing interest rates.
- The Summary of Key Results section should be just that—*key* results. These outputs help you make good business decisions. Key results frequently include net income before taxes (profit) and end-of-year cash on hand (how much cash your business has). However, if you are creating a DSS model on alternative capital projects, your key results can also include cost savings, net present value of a project, or rate of return for an investment.
- The Constants section holds known values needed to perform other calculations. You use a Constants section rather than just including the values in formulas so that you can input new values if they change. This approach makes your DSS model more flexible.

AT THE KEYBOARD

Enter the Excel skeleton shown in Figures C-1 and C-2, or use the skeleton from your data files.

NOTE

When you see NA (Not Applicable) in a cell, do not enter any values or formulas in the cell. The cells that contain values in the 2015 column are used by other cells for calculations. In this example, you are mainly interested in what happens in 2016 and 2017. The rest of the cells are "Not Applicable."

Filling in the "Easy" Formulas

The next step in building a spreadsheet is to fill in the "easy" formulas. To begin, format all the cells that will contain monetary values as Currency with zero decimal places:

- Constants—cell C5
- Summary of Key Results—C12 to C15, D12 to D15
- Calculations (Expansion)—C18, C19, D18, D19
- Calculations (No Expansion)—C24, C25, D24, D25
- Income and Cash Flow Statements (Expansion)—B36, C29 to C36, D29 to D36
- Income and Cash Flow Statements (No Expansion)—B45, C39 to C45, D39 to D45

NOTE

With the insertion point in cell C12 (where the $0 appears), note the editing window—the white space at the top of the spreadsheet to the right of the f_x symbol. The cell's contents, whether it is a formula or value, should appear in the editing window. In this case, the window shows =C35.

The Summary of Key Results section (see Figure C-3) will contain the values you calculate in the Income and Cash Flow Statements sections. To copy the cell contents for this section, move your cursor to cell C12, click the cell, type =C35, and press Enter. If you formatted your money cells properly, a $0 should appear in cell C12.

C12	fx	=C35		
	A	B	C	D
		2015	2016	2017
11	**Summary of Key Results**	2015	2016	2017
12	Net Income after Taxes (Expansion)	NA	$0	
13	End-of-year Cash on Hand (Expansion)	NA		
14	Net Income after Taxes (No Expansion)	NA		
15	End-of-year Cash on Hand (No Expansion)	NA		

Source: Microsoft product screenshots used with permission from Microsoft Corporation

FIGURE C-3   Value from cell C35 (Net Income after Taxes) copied to cell C12

Because cell C35 does not contain a value yet, Excel assumes that the empty cell has a numerical value of 0. When you put a formula in cell C35 later, cell C12 will echo the resulting answer. Because Net Income after Taxes (Expansion) for 2017 (cell D35) and its corresponding cell in Summary of Key Results (cell D12) are both directly to the right of the values for 2016, you can either type =D35 into cell D12 or copy cell C12 to D12. Excel allows more than one way to perform the copy operation; here is one method:

1. Click in a cell or click and drag to select the range of cells you want to copy.
2. Hold down the Control key and press C (Ctrl+C).
3. A moving dashed box called a *marquee* should now be animated over the cell(s) selected for copying.
4. Select the cell(s) where you want to copy the data.
5. Hold down the Control key and press V (Ctrl+V). Cell D12 should now contain $0, but actually it has a reference to cell D35. Click cell D12 and look again at the editing window; it should display =D35.

The same thinking applies to End of Year Cash on Hand (Expansion). Enter =C36 into cell C13, then copy C13 to D13. Cells C14, C15, D14, and D15 represent Net Income after Taxes and End-of-year Cash on Hand for both years of No Expansion; these cells are mirrors of cells C44, C45, D44, and D45 in the last section. Select cell C14, type =C44, and press Enter. Select cell C14 again, use the Copy command, and paste the contents into cell D14 (see Figure C-4).

		2015	2016	2017
11	**Summary of Key Results**	2015	2016	2017
12	Net Income after Taxes (Expansion)	NA	$0	$0
13	End-of-year Cash on Hand (Expansion)	NA	$0	$0
14	Net Income after Taxes (No Expansion)	NA	$0	
15	End-of-year Cash on Hand (No Expansion)	NA		
16				
17	**Calculations (Expansion)**	2015	2016	2017
18	Total Sales Dollars	$350,000		
19	Cost of Goods Sold	$245,000		
20	Cost of Goods Sold (as a percent of Sales)	70%		
21	Interest Rate for Business Loan		NA	NA

Source: Microsoft product screenshots used with permission from Microsoft Corporation

FIGURE C-4   Copying the formula from cell C14 to cell D14

Because Excel uses *relative* cell references by default, copying cell C14 into cell D14 will copy and paste the contents of cell D44 (the cell adjacent to C44) into cell D14. See Figure C-5.

D14	fx	=D44		
	A	B	C	D
		2015	2016	2017
11	**Summary of Key Results**	2015	2016	2017
12	Net Income after Taxes (Expansion)	NA	$0	$0
13	End-of-year Cash on Hand (Expansion)	NA	$0	$0
14	Net Income after Taxes (No Expansion)	NA	$0	$0
15	End-of-year Cash on Hand (No Expansion)	NA		

Source: Microsoft product screenshots used with permission from Microsoft Corporation

FIGURE C-5   Formula from cell D44 pasted into cell D14

Use the Copy command again, this time downward from cells C14 and D14, to complete cells C15 and D15. If you are successful, the formula in the editing window for cell C15 will be "=C45" and for cell D15 will display "=D45."

You will create the formulas for the two Calculations sections last because they are the hardest formulas. Next, you will create the formulas for the two Income and Cash Flow Statements sections; all the cells in these two sections should be formatted as Currency with zero decimal places.

As shown in Figure C-6, the Beginning-of-year Cash on Hand for 2016 is the End-of-year Cash on Hand for 2015. In cell C29, type =B36. A handy shortcut is to type the "=" sign, immediately move your mouse pointer to the cell you want to designate, and then click the left mouse button. Excel will enter the cell location into the formula for you. This shortcut is especially useful if you want to avoid making a typing error.

PMT			f_x	=B36			
		A		B	C	D	
28	**Income and Cash Flow Statements (Expansion)**			**2015**	**2016**	**2017**	
29	Beginning-of-year Cash on Hand			NA	=B36		
30	Sales (Revenue)			NA			
31	Cost of Goods Sold			NA			
32	Business Loan Payment			NA			
33	Income before Taxes			NA			
34	Income Tax Expense			NA			
35	Net Income after Taxes			NA			
36	End-of-year Cash on Hand			$15,000			
37							
38	**Income and Cash Flow Statements (No Expansion)**			**2015**	**2016**	**2017**	
39	Beginning-of-year Cash on Hand			NA			
40	Sales (Revenue)			NA			
41	Cost of Goods Sold			NA			
42	Income before Taxes			NA			
43	Income Tax Expense			NA			
44	Net Income after Taxes			NA			
45	End-of-year Cash on Hand			$15,000			

Source: Microsoft product screenshots used with permission from Microsoft Corporation

FIGURE C-6   End-of-year Cash on Hand for 2015 copied to Beginning-of-year Cash on Hand for 2016

Likewise, copy the other three End-of-year Cash on Hand cells to the Beginning-of-year Cash on Hand cells for both Income and Cash Flow Statements (cells D29, C39, and D39).

The Sales (Revenue) cells C30, D30, C40, and D40 are simply copies of cells C18, D18, C24, and D24, respectively, from the Calculations sections (both Expansion and No Expansion). Note that all four cells will display $0 until you enter the formulas in the Calculations sections (see Figure C-7).

D40			f_x	=D24			
		A		B	C	D	
23	**Calculations (No Expansion)**			**2015**	**2016**	**2017**	
24	Total Sales Dollars			$350,000			
25	Cost of Goods Sold			$245,000			
26	Cost of Goods Sold (as a percent of Sales)			70%			
27							
28	**Income and Cash Flow Statements (Expansion)**			**2015**	**2016**	**2017**	
29	Beginning-of-year Cash on Hand			NA	$15,000	$0	
30	Sales (Revenue)			NA	$0	$0	
31	Cost of Goods Sold			NA			
32	Business Loan Payment			NA			
33	Income before Taxes			NA			
34	Income Tax Expense			NA			
35	Net Income after Taxes			NA			
36	End-of-year Cash on Hand			$15,000			
37							
38	**Income and Cash Flow Statements (No Expansion)**			**2015**	**2016**	**2017**	
39	Beginning-of-year Cash on Hand			NA	$15,000	$0	
40	Sales (Revenue)			NA	$0	$0	
41	Cost of Goods Sold			NA			
42	Income before Taxes			NA			
43	Income Tax Expense			NA			
44	Net Income after Taxes			NA			
45	End-of-year Cash on Hand			$15,000			

Source: Microsoft product screenshots used with permission from Microsoft Corporation

FIGURE C-7   Sales Revenue cells copied from the Calculations sections

The Cost of Goods Sold cells C31, D31, C41, and D41 are simply copies of the contents of cells C19, D19, C25, and D25, respectively, from the Calculations sections. This is shown in Figure C-8.

D41			f_x	=D25							
	A		B	C	D	E	F	G	H		
23	**Calculations (No Expansion)**		**2015**	**2016**	**2017**						
24	Total Sales Dollars		$350,000								
25	Cost of Goods Sold		$245,000								
26	Cost of Goods Sold (as a percent of Sales)		70%								
27											
28	**Income and Cash Flow Statements (Expansion)**		**2015**	**2016**	**2017**						
29	Beginning-of-year Cash on Hand		NA	$15,000	$0						
30	Sales (Revenue)		NA	$0	$0						
31	Cost of Goods Sold		NA	$0	$0						
32	*Business Loan Payment*		NA								
33	Income before Taxes		NA								
34	Income Tax Expense		NA								
35	Net Income after Taxes		NA								
36	End-of-year Cash on Hand		$15,000								
37											
38	**Income and Cash Flow Statements (No Expansion)**		**2015**	**2016**	**2017**						
39	Beginning-of-year Cash on Hand		NA	$15,000	$0						
40	Sales (Revenue)		NA	$0	$0						
41	Cost of Goods Sold		NA	$0	$0						
42	Income before Taxes		NA								
43	Income Tax Expense		NA								
44	Net Income after Taxes		NA								
45	End-of-year Cash on Hand		$15,000								
46											

Source: Microsoft product screenshots used with permission from Microsoft Corporation

FIGURE C-8   Cost of Goods Sold cells copied from the Calculations sections

Next you determine the Business Loan Payment for cells C32 and D32—notice that it is only present in the Income and Cash Flow Statements (Expansion) section, because if you do not expand the business, you do not need the business loan of $100,000. Excel has financial formulas to figure out loan payments. To determine a loan payment, you need to know three things: the amount being borrowed (cell C5 in the Constants section), the interest rate (cell B21 in the Calculations-Expansion section), and the number of payment periods. At the beginning of the tutorial, you learned that the bank was willing to loan money at either 5% or 4% interest compounded annually, to be paid over 10 years. Normally, banks require businesses to make monthly payments on their loans and compound the interest monthly, in which case you would enter 120 (12 months/year × 10 years) for the number of payments and divide the annual interest rate by 12 to get the period interest rate. This formula is important to remember when you enter the business world, but for now you will simplify the calculation by specifying one loan payment per year compounded annually. To put in the payment formula, click cell C32, then click the f_x symbol next to the editing window. The Payment function is called PMT, so type PMT in the Insert Function window and click Go—you will see a short description of the function with its arguments, as shown in Figure C-9.

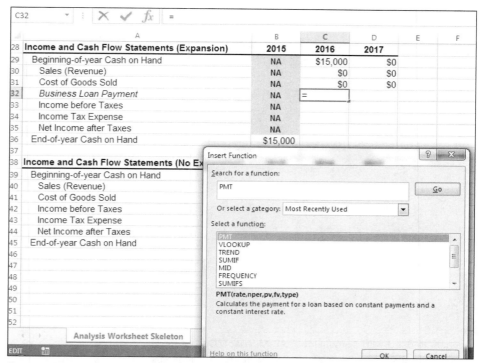

Source: Microsoft product screenshots used with permission from Microsoft Corporation

FIGURE C-9   Accessing the PMT function in Excel for cell C32

> **NOTE**
>
> Rate is the interest rate per period of the loan, Nper is an abbreviation for the number of loan periods, and Pv is an abbreviation for Present Value, the amount of money you are borrowing "today." The PMT function can determine a series of equal loan payments necessary to pay back the amount borrowed, plus the accumulated compound interest over the life of the loan.

When you click OK, the resulting window allows you to enter the cells or values needed in the function arguments (see Figure C-10). In the Rate text box, enter B21, which is the cell that will contain the calculated interest rate. In the Nper text box, enter 10 (for 10 years). In the Pv text box, enter C5, which is the cell that contains the loan amount.

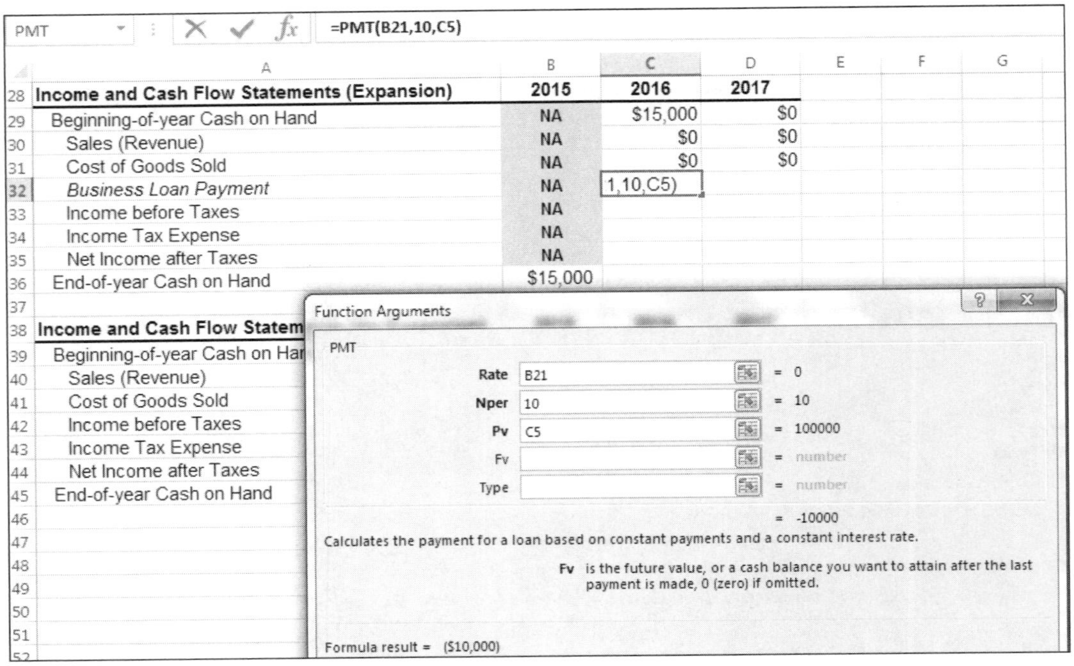

FIGURE C-10   The Function Arguments window for the PMT function with the values filled in

> **NOTE**
>
> Be careful if you decide to copy the PMT formula from cell C32 into cell D32, because the Copy command will change the cells in the formula arguments to the next adjacent cells. To make the Copy command work correctly, you have two options. First, you can change the Rate and Pv cells in the cell C32 formula from *relative reference* (B21, C5) to *absolute reference* ($B$21, $C$5). Your other option is to re-insert the PMT function into cell D32 and type the same arguments as before in the boxes. Absolute referencing of a cell (using $ signs in front of the Column and Row designators) "anchors" the cell so that when the Copy command is used, the destination cell will refer back to the same cells that the source cell used. If necessary, consult the Excel online Help for an explanation of relative and absolute cell references.

When you click OK ($10,000) should appear in cell C32. Payments in Excel always appear as negative numbers, which is why the number has parentheses around it. (Depending on your cell formatting, the number may also appear in red.) Next, you need to have the same payment amount in cell D32 (for 2017). Because the PMT function creates equal payments over the life of the loan, you can simply type =C32 into cell D32.

The next line in the Income and Cash Flow Statements is Income before Taxes, which is an easy calculation. It is the Sales minus the Cost of Goods Sold, minus the Business Loan Payment. However, because the PMT function shows the loan payment as a negative number, you will instead add the Business Loan Payment. In cell C33, enter =C30-C31+C32. Again, a negative $10,000 should be displayed, as the cells other than the loan payment currently have zero in them. Copy cell C33 to cell D33. In cell C42 of the next section below (No Expansion), enter =C40-C41. (There is no loan payment in this section to put in the calculation.) Next, copy cell C42 to cell D42. At this point, your Income and Cash Flow Statements should look like Figure C-11.

D42   =D40-D41

A	B	C	D
28 Income and Cash Flow Statements (Expansion)	2015	2016	2017
29 Beginning-of-year Cash on Hand	NA	$15,000	$0
30 Sales (Revenue)	NA	$0	$0
31 Cost of Goods Sold	NA	$0	$0
32 Business Loan Payment	NA	($10,000)	($10,000)
33 Income before Taxes	NA	-$10,000	-$10,000
34 Income Tax Expense	NA		
35 Net Income after Taxes	NA		
36 End-of-year Cash on Hand	$15,000		
37			
38 Income and Cash Flow Statements (No Expansion)	2015	2016	2017
39 Beginning-of-year Cash on Hand	NA	$15,000	$0
40 Sales (Revenue)	NA	$0	$0
41 Cost of Goods Sold	NA	$0	$0
42 Income before Taxes	NA	$0	$0
43 Income Tax Expense	NA		
44 Net Income after Taxes	NA		
45 End-of-year Cash on Hand	$15,000		

Source: Microsoft product screenshots used with permission from Microsoft Corporation

FIGURE C-11   The Income and Cash Flow Statements completed up to Income before Taxes

Income Tax Expense is the most complex formula for these sections. Because you do not pay income tax when you have no income or a loss, you must use a formula that enters 0 if there is no income or a loss, or calculates the tax rate on a positive income. You can use the IF function in Excel to enter one of two results in a cell, depending on whether a defined logical statement is true or false. One way to create an IF function is to select cell C34, then click the f_x symbol next to the cell editing window. When the Insert Function window appears, as shown in Figure C-12, type IF in the "Search for a function" text box, and click the Go button if necessary. The IF function should appear. When you click OK, the Function Arguments window appears. You could then "fill in the blanks" to create the IF statement, in the same way you did for the PMT function.

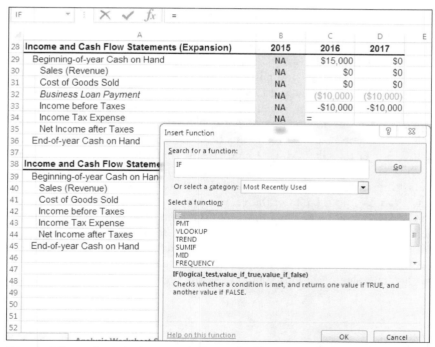

Source: Microsoft product screenshots used with permission from Microsoft Corporation

FIGURE C-12   The IF function

However, the logic of an IF statement is usually clear enough, so it is often just as easy to enter the full statement. The syntax of an IF statement is: *IF (test condition, what Excel should do if the test condition is True, what Excel should do if the test condition is False)*. In this case, you want to test whether income before taxes is less than or equal to zero. If the results of the test are True (income before taxes is less than or equal to zero), Excel should enter a zero for tax expense. Otherwise, taxes are equal to income before taxes multiplied by the tax rate. In Figure C-13, the 2016 income before taxes is shown in cell C33. The formula for income tax expense is included for cell C34.

C34			× ✓ f_x	=IF(C33<=0,0,C33*C4)			
	A			B	C	D	
				2015	2016	2017	
28	Income and Cash Flow Statements (Expansion)						
29	Beginning-of-year Cash on Hand			NA	$15,000	$0	
30	Sales (Revenue)			NA	$0	$0	
31	Cost of Goods Sold			NA	$0	$0	
32	*Business Loan Payment*			NA	($10,000)	($10,000)	
33	Income before Taxes			NA	-$10,000	-$10,000	
34	Income Tax Expense			NA	$0		
35	Net Income after Taxes			NA			
36	End-of-year Cash on Hand			$15,000			
37							
38	Income and Cash Flow Statements (No Expansion)			2015	2016	2017	
39	Beginning-of-year Cash on Hand			NA	$15,000	$0	
40	Sales (Revenue)			NA	$0	$0	
41	Cost of Goods Sold			NA	$0	$0	
42	Income before Taxes			NA	$0	$0	
43	Income Tax Expense			NA			
44	Net Income after Taxes			NA			
45	End-of-year Cash on Hand			$15,000			
46							
47							
48							
49							
50							
51							
52							

Source: Microsoft product screenshots used with permission from Microsoft Corporation

FIGURE C-13   IF statement entered into cell C34

Because you had negative income, the cell should display a zero for now. Because the same logic applies to Net Income After Taxes in 2017, you can simply copy and paste the formula from cell C34 to cell D34. You also have to calculate income tax expense for the Income and Cash Flow Statements (No Expansion). As a test of your understanding, enter the proper IF statement in cell C43. Again, the cell will display $0 for an answer. Copy cell C43 to cell D43 to complete the Income Tax Expense line for No Expansion.

Net Income after Taxes is simply the Income before Taxes minus the Income Tax Expense. Enter the formula into cell C35, then copy cell C35 over to cells D35, C44, and D44. If you did this correctly, cells C35 and D35 will display a negative $10,000, and cells C44 and D44 will display $0.

End-of-year Cash on Hand, the last line in both Income and Cash Flow Statements sections, is not difficult either. Conceptually, the cash you have at the end of the year is equal to your Beginning-of-year Cash on Hand plus your Net Income after Taxes. Enter the formula into cell C36, then copy cell C36 over to cell D36. Note that because the Income and Cash Flow Statements (No Expansion) do not have a line item for Business Loan Payment, you cannot copy the same command down to it. You have to enter the formula manually for cell C45, which is =C39+C44. However, you can copy cell C45 to cell D45 to finish the Income and Cash Flow Statements sections. The completed sections should look like Figure C-14.

D45	∣ ✕ ✓ ƒx	=D39+D44		

	A	B	C	D
28	**Income and Cash Flow Statements (Expansion)**	**2015**	**2016**	**2017**
29	Beginning-of-year Cash on Hand	NA	$15,000	$5,000
30	Sales (Revenue)	NA	$0	$0
31	Cost of Goods Sold	NA	$0	$0
32	*Business Loan Payment*	NA	($10,000)	($10,000)
33	Income before Taxes	NA	-$10,000	-$10,000
34	Income Tax Expense	NA	$0	$0
35	Net Income after Taxes	NA	-$10,000	-$10,000
36	End-of-year Cash on Hand	$15,000	$5,000	-$5,000
37				
38	**Income and Cash Flow Statements (No Expansion)**	**2015**	**2016**	**2017**
39	Beginning-of-year Cash on Hand	NA	$15,000	$15,000
40	Sales (Revenue)	NA	$0	$0
41	Cost of Goods Sold	NA	$0	$0
42	Income before Taxes	NA	$0	$0
43	Income Tax Expense	NA	$0	$0
44	Net Income after Taxes	NA	$0	$0
45	End-of-year Cash on Hand	$15,000	$15,000	$15,000

Source: Microsoft product screenshots used with permission from Microsoft Corporation

FIGURE C-14   The completed Income and Cash Flow Statements sections

Filling in the "Hard" Formulas

To finish the spreadsheet, you will enter values in the Inputs section and write the formulas in both Calculations sections.

AT THE KEYBOARD

In cell C8, enter an R for Recession, and in cell C9, enter H for High Inflation. You could enter any values here, but these two values will work with the IF functions you will write later. Recall that you did not use separate inputs for 2016 and 2017. You are assuming that the economic outlook or inflation rate that exists for 2016 will extend into 2017. However, because you are using the same inputs from these two locations, you must remember to use *absolute* cell references to both cells C8 and C9 in the various IF statements if you want to use a Copy command for adjacent cells. Your Inputs section should look like the one in Figure C-15.

	A	B	C	D
7	**Inputs**	**2015**	**2016**	**2017**
8	Economic Outlook (R=Recession, B=Boom)	NA	R	NA
9	Inflation Outlook (H=High, L=Low)	NA	H	NA

Source: Microsoft product screenshots used with permission from Microsoft Corporation

FIGURE C-15   The Inputs section with values entered in cells C8 and C9

Remember that you referred to cell addresses in both Calculations sections in your formulas in the Income and Cash Flow Statements sections. Now you will enter formulas for these calculations. If necessary, format the four Total Sales Dollars cells and the four Cost of Goods Sold cells in the Calculations sections as Currency with no decimal places.

As described at the beginning of the tutorial, the forecast for Total Sales Dollars is a function of both the Economic Outlook and whether you expand the business. The following table lists the predicted sales growth percentages:

Sales Growth Forecast—Collegetown Thrift Shop

	Business Expansion	No Business Expansion
Recession-R	30%	20%
Boom-B	15%	5%

You will use IF formulas to forecast Total Sales Dollars. In cell C18, you will need an IF statement that has the following logic:

Logical_test: $C$8="R" (Note that you must use absolute cell referencing for cell C8 and quotation marks for Excel to recognize a text string.)

Value_if_test_is_true: B18*1.3 (2015 sales multiplied by 1.3 for 30% sales growth)

Value_if_test_is_false: B18*1.15 (2015 sales multiplied by 1.15 for 15% sales growth)

Figure C-16 shows the needed IF statement.

C18	f_x =IF($C$8="R",B18*1.3,B18*1.15)			
	A	B	C	D
7	Inputs	2015	2016	2017
8	Economic Outlook (R=Recession, B=Boom)	NA	R	NA
9	Inflation Outlook (H=High, L=Low)	NA	H	NA
10				
11	Summary of Key Results	2015	2016	2017
12	Net Income after Taxes (Expansion)	NA	$298,150	-$10,000
13	End-of-year Cash on Hand (Expansion)	NA	$313,150	$303,150
14	Net Income after Taxes (No Expansion)	NA	$0	$0
15	End-of-year Cash on Hand (No Expansion)	NA	$15,000	$15,000
16				
17	Calculations (Expansion)	2015	2016	2017
18	Total Sales Dollars	$350,000	$455,000	
19	Cost of Goods Sold	$245,000		
20	Cost of Goods Sold (as a percent of Sales)	70%		
21	Interest Rate for Business Loan		NA	NA
22				
23	Calculations (No Expansion)	2015	2016	2017
24	Total Sales Dollars	$350,000		
25	Cost of Goods Sold	$245,000		
26	Cost of Goods Sold (as a percent of Sales)	70%		

Source: Microsoft product screenshots used with permission from Microsoft Corporation

FIGURE C-16   Using an IF statement to enter the Total Sales Dollars forecast for 2016

Cell C18 displays $455,000 because 30% of $350,000 is $105,000, and $350,000 plus $105,000 equals $455,000. (So, it appears that the formula returned a "true" value with an R inserted in cell C8.) Because you "anchored" cell C8 by entering $C$8, copy this formula over to cell D18 for the year 2017.

Once you complete the Total Sales Dollars cells for the Expansion scenario, go down to the Calculations (No Expansion) section and use IF statements to enter formulas for the Total Sales Dollars. Use a 20% sales growth factor for Recession and 5% for Boom. You can copy the formula from cell C18 into cell C24, but you then will have to use the editing window to change the values in the true and false arguments from 1.3 and 1.15 to 1.2 and 1.05, respectively, to reflect the fact that you did not expand the business. See Figure C-17.

C24	f_x =IF($C$8="R",B24*1.2,B24*1.05)			
	A	B	C	D
17	Calculations (Expansion)	2015	2016	2017
18	Total Sales Dollars	$350,000	$455,000	$591,500
19	Cost of Goods Sold	$245,000		
20	Cost of Goods Sold (as a percent of Sales)	70%		
21	Interest Rate for Business Loan		NA	NA
22				
23	Calculations (No Expansion)	2015	2016	2017
24	Total Sales Dollars	$350,000	$420,000	
25	Cost of Goods Sold	$245,000		
26	Cost of Goods Sold (as a percent of Sales)	70%		
27				
28	Income and Cash Flow Statements (Expansion)	2015	2016	2017
29	Beginning-of-year Cash on Hand	NA	$15,000	$311,173

Source: Microsoft product screenshots used with permission from Microsoft Corporation

FIGURE C-17   Total Sales Dollars in the No Expansion scenario

As before, you can now copy cell C24 to cell D24. You have completed the Total Sales Dollars calculations.

The Cost of Goods Sold (cells C19, D19, C25, and D25) is the Total Sales Dollars multiplied by the Cost of Goods Sold as a percent of Sales. In cell C19, type =C18*C20 and press Enter. Copy cell C19 and paste the contents into cells D19, C25, and D25. Your answers will be $0 until you enter the formulas for the Cost of Goods Sold as a percent of Sales.

The Cost of Goods Sold as a percent of Sales (cells C20, D20, C26, and D26) was 70% in 2015. In variety merchandising for resold items, it is easier to use an aggregate measure such as Cost of Goods Sold as a percent of Sales rather than trying to capture an individual Cost of Goods Sold for each item. From the 2015 data, you determined that for every dollar of sales you collected in 2015, you spent 70 cents purchasing the item and preparing it for resale. You will use that percentage as a basis for forecasting Cost of Goods Sold as a percent of Sales, applying an appropriate inflation factor for the cost of acquiring the stock for sale. The following table lists the predicted inflation percentages for Cost of Goods Sold.

Cost of Goods Sold Forecast—Collegetown Thrift Shop

	Business Expansion	No Business Expansion
High Inflation	6%	6%
Low Inflation	2%	2%

As with Total Sales Dollars previously, you will again use an IF statement to calculate the Cost of Goods Sold as a percent of Sales. In cell C20, type the following:

=IF($C$9="H",B20*1.06,B20*1.02)

This expression means that if the text string in cell C9 is the letter H, you multiply the value in cell B20 by 1.06 (6% inflation). If the value in cell C9 is not an H, multiply the value in cell B20 by 1.02 (2% inflation). The value in cell B20 was the baseline Cost of Goods Sold as a percent of Sales in 2015, which was 70%. You can now copy cell C20 and paste the contents into cell D20.

Because the inflation percentages were exactly the same for both the Expansion and No Expansion calculations, you can also copy cell C20 and paste the contents into cells C26 and D26. Your Calculations sections should now look like Figure C-18.

D26	fx	=IF($C$9="H",C26*1.06,C26*1.02)		
	A	B	C	D
17	**Calculations (Expansion)**	**2015**	**2016**	**2017**
18	Total Sales Dollars	$350,000	$455,000	$591,500
19	Cost of Goods Sold	$245,000	$337,610	$465,227
20	Cost of Goods Sold (as a percent of Sales)	70%	74%	79%
21	Interest Rate for Business Loan		NA	NA
22				
23	**Calculations (No Expansion)**	**2015**	**2016**	**2017**
24	Total Sales Dollars	$350,000	$420,000	$504,000
25	Cost of Goods Sold	$245,000	$311,640	$396,406
26	Cost of Goods Sold (as a percent of Sales)	70%	74%	79%

Source: Microsoft product screenshots used with permission from Microsoft Corporation

FIGURE C-18 Calculations sections nearly complete

The last item in the Calculations section is the Interest Rate for Business Loan (cell B21). Remember the bank's statement that if the economy recovers, it could lower the interest rate from 5% to 4%. So, you will need one more IF statement to insert into cell B21 based on the economic outlook. If the economic outlook is for a Recession (R), then the interest rate will be 5% annually; if the outlook is for a Boom (B), then the interest rate will be 4% annually. Click cell B21, type =IF(C8="R",.05,.04), and press Enter.

You will immediately notice that 5% appears in the cell because you have R in the input cell for Economic Outlook. You may also notice that you now have a negative $12,950 in the Business Loan Payment cells (C32 and D32). See Figure C-19 to compare your results.

28	Income and Cash Flow Statements (Expansion)	2015	2016	2017
29	Beginning-of-year Cash on Hand	NA	$15,000	$84,974
30	Sales (Revenue)	NA	$455,000	$591,500
31	Cost of Goods Sold	NA	$337,610	$465,227
32	*Business Loan Payment*	NA	($12,950)	($12,950)
33	Income before Taxes	NA	$104,440	$113,323
34	Income Tax Expense	NA	$34,465	$39,663
35	Net Income after Taxes	NA	$69,974	$73,660
36	End-of-year Cash on Hand	$15,000	$84,974	$158,634
37				
38	Income and Cash Flow Statements (No Expansion)	2015	2016	2017
39	Beginning-of-year Cash on Hand	NA	$15,000	$87,601
40	Sales (Revenue)	NA	$420,000	$504,000
41	Cost of Goods Sold	NA	$311,640	$396,406
42	Income before Taxes	NA	$108,360	$107,594
43	Income Tax Expense	NA	$35,759	$37,658
44	Net Income after Taxes	NA	$72,601	$69,936
45	End-of-year Cash on Hand	$15,000	$87,601	$157,537

Source: Microsoft product screenshots used with permission from Microsoft Corporation

FIGURE C-19 Income and Cash Flow Statements in finished spreadsheet

You can change the economic inputs in four different combinations: R-H, R-L, B-H, B-L. This allows you to see the impact on your net income and cash on hand both for expanding and not expanding in the Summary of Key Results section. However, you have another more powerful way to display scenario results. In the next section, you will learn how to tabulate the financial results of the four possible combinations using an Excel tool called Scenario Manager.

SCENARIO MANAGER

You are now ready to evaluate the four possible outcomes for your DSS model. Because this is a simple, four-outcome model, you could have created four different spreadsheets, one for each set of outcomes, and then transferred the financial information from each spreadsheet to a Summary Report.

In essence, Scenario Manager performs the same task. It runs the model for all the requested outcomes and presents a tabular summary of the results. This summary is especially useful for reports and presentations needed by upper managers, financial investors, or in this case, the bank.

To review, the four possible combinations of input values are: R-H (Recession and High Inflation), R-L (Recession and Low Inflation), B-H (Boom and High Inflation), and B-L (Boom and Low Inflation). You could consider each combination of inputs a separate scenario. For each of these scenarios, you are interested in four outputs: Net Income after Taxes for Expansion and No Expansion, and End-of-year Cash on Hand for Expansion and No Expansion.

Scenario Manager runs each set of combinations and then records the specified outputs as a summary into a separate worksheet. You can use these summary values as a table of numbers and print it, or you can copy them into a Microsoft Word document or a PowerPoint presentation. You can also use the data table to build a chart or graph, which you can put into a report or presentation.

When you define a scenario in Scenario Manager, you name it and identify the input cells and input values. Then you identify the output cells so Scenario Manager can capture the outputs in a summary sheet.

AT THE KEYBOARD

To start, click the Data tab on the Ribbon. In the Data Tools group, click the What-If Analysis button, then note that Scenario Manager appears as an option (see Figure C-20).

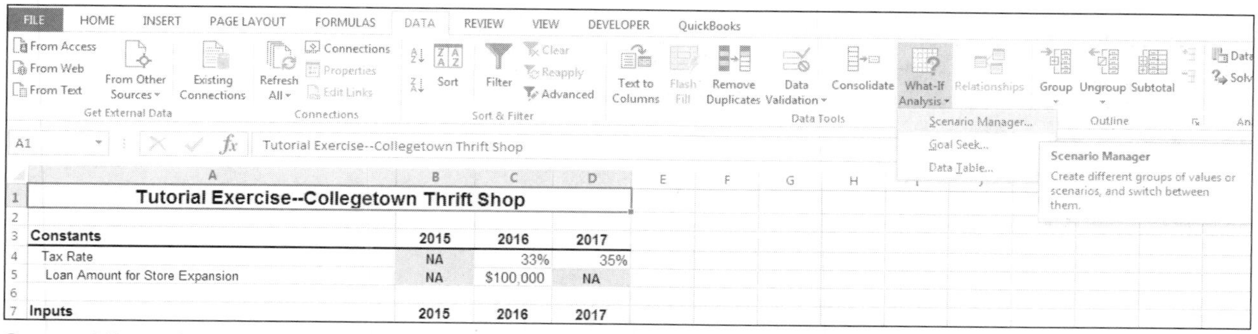

Source: Microsoft product screenshots used with permission from Microsoft Corporation

FIGURE C-20   Scenario Manager option in the What-If Analysis menu

Select Scenario Manager in the menu. The Scenario Manager window appears (see Figure C-21), but no scenarios are defined. Use the window to add, delete, or edit scenarios.

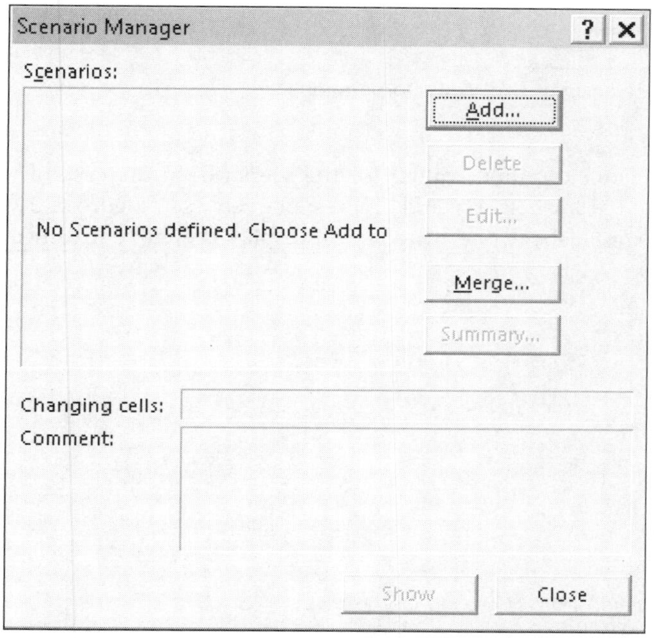

Source: Microsoft product screenshots used with permission from Microsoft Corporation

FIGURE C-21   Initial Scenario Manager window

NOTE

When working with the Scenario Manager window and any following windows, do not use the Enter key to navigate. Use mouse clicks to move from one step to the next.

To define a scenario, click the Add button. The Edit Scenario window appears. (The title bar displays Add Scenario until you enter data in the fields.) Enter Recession-High Inflation in the field under Scenario name. Then type the input cells in the Changing cells field (in this case, C8:C9). You can use the button next to the field to select the changing cells in your spreadsheet. If you do, Scenario Manager changes the cell references to absolute cell references, which is acceptable (see Figure C-22).

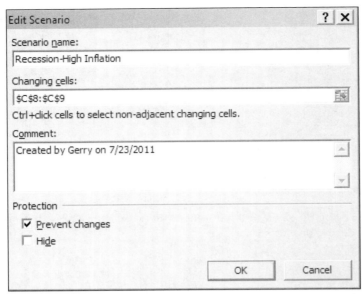

Source: Microsoft product screenshots used with permission from Microsoft Corporation

FIGURE C-22   Defining a scenario name and input cells

Click OK to open the Scenario Values window. Enter the input values for the scenario. In the case of Recession and High Inflation, the values will be R and H for cells C8 and C9, respectively (see Figure C-23). Note that if you already have entered values in the spreadsheet, the window will display the current values. Make sure to enter the correct values.

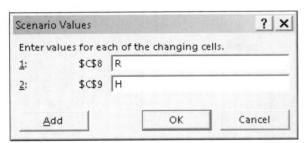

Source: Microsoft product screenshots used with permission from Microsoft Corporation

FIGURE C-23   Entering values for the scenario's input cells

Click OK to return to the Scenario Manager window. In the same way, enter the other three scenarios: Recession-Low Inflation, Boom-High Inflation, and Boom-Low Inflation (R-L, B-H, and B-L), and their related input values. When you finish, you should see the names and changing cells for the four scenarios (see Figure C-24).

Source: Microsoft product screenshots used with permission from Microsoft Corporation

FIGURE C-24   Scenario Manager window with all four scenarios entered

You can now create a summary sheet that displays the results of running the four scenarios. Click the Summary button to open the Scenario Summary window, as shown in Figure C-25. You must now enter the output cell addresses in Excel—they will be the same for all four scenarios. Recall that you created a section in your spreadsheet called Summary of Key Results. You are primarily interested in the results at the end of 2017, so you will choose the four cells that represent the Net Income after Taxes and End-of-year Cash on Hand, and then use them for both the expansion scenario and the non-expansion scenario. These cells are D12 to D15 in your spreadsheet. Either type D12:D15 or use the button next to the Result cells field and select those cells in the spreadsheet.

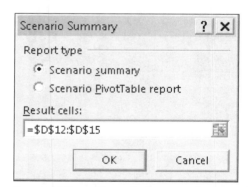

Source: Microsoft product screenshots used with permission from Microsoft Corporation

FIGURE C-25   Scenario Summary window with Result cells entered

Another good reason for having a Summary of Key Results section is that it provides a contiguous range of cells to define for summary output. However, if you want to add output from other cells in the spreadsheet, simply separate each cell or range of cells in the window with a comma. Next, click OK. Excel runs each set of inputs in the background, collects the results from the result cells, and then enters the data in a new sheet called Scenario Summary (the name on the sheet's lower tab), as shown in Figure C-26.

	A	B	C	D	E	F	G	H
1								
2		Scenario Summary						
3				Current Values:	Recession-High Inflation	Recession-Low Inflation	Boom-High Inflation	Boom-Low Inflation
5		Changing Cells:						
6		$C$8	R	R	R	B	B	
7		$C$9	H	H	L	H	L	
8		Result Cells:						
9		$D$12		$73,660	$73,660	$96,052	$56,216	$73,738
10		$D$13		$158,634	$158,634	$189,562	$132,531	$157,605
11		$D$14		$69,936	$69,936	$89,015	$53,545	$68,152
12		$D$15		$157,537	$157,537	$184,496	$132,071	$153,573
13		Notes: Current Values column represents values of changing cells at						
14		time Scenario Summary Report was created. Changing cells for each						
15		scenario are highlighted in gray.						

Source: Microsoft product screenshots used with permission from Microsoft Corporation

FIGURE C-26   Scenario Summary sheet created by Scenario Manager

As you can see, the output created by the Scenario Summary sheet is not formatted for easy reading. You do not know which results are the net income and cash on hand, and you do not know which results are for Expansion vs. No Expansion, because Scenario Manager listed only the cell addresses. Scenario Manager also listed a separate column (column D) for the current input values in the spreadsheet, which are the same as the values in column E. It also left a blank column (column A) in the spreadsheet.

Fortunately, it is fairly easy to format the output. Delete columns D and A, put in the labels for cell addresses in the new column A, and then retitle the Scenario Summary as Collegetown Thrift Shop Financial Forecast, End of Year 2017 (because you are looking only at Year 2017 results). You can also make the results columns narrower by breaking the column headings into two lines; place your cursor in the editing window where you want to break the words, and then press Alt+Enter. Add a heading for column B (Cell Address). Create the title, and center the column headings and the input cell values (R, B, H, and L). Leave your financial data right-justified to keep the numbers lined up correctly. Finally, delete the notes in Rows 13 through 15—they are no longer needed. Figure C-27 shows a formatted Scenario Summary worksheet.

	A	B	C	D	E	F
1						
2	Scenario Summary -- Collegetown Thrift Shop Financial Forecast -- End of Year 2017					
3		Cell Address	Recession-High Inflation	Recession-Low Inflation	Boom-High Inflation	Boom-Low Inflation
5	Changing Cells:					
6	Economic Outlook: R-Recession, B-Boom	$C$8	R	R	B	B
7	Inflation: H-High, L-Low	$C$9	H	L	H	L
8	Result Cells:					
9	Net Income After Taxes-Expansion	$D$12	$73,660	$96,052	$56,216	$73,738
10	End-Of-Year Cash On Hand-Expansion	$D$13	$158,634	$189,562	$132,531	$157,605
11	Net Income After Taxes- No Expansion	$D$14	$69,936	$89,015	$53,545	$68,152
12	End-Of-Year Cash On Hand-No Expansion	$D$15	$157,537	$184,496	$132,071	$153,573

Source: Microsoft product screenshots used with permission from Microsoft Corporation

FIGURE C-27   Scenario Summary worksheet after formatting

Interpreting the Results

Now that you have good data, what do you do with it? Remember, you wanted to see if taking a $100,000 business loan to expand the thrift shop was a good financial decision. The shop's success so far ($350,000 of sales in 2015) would seem to make expansion a good risk. But how good a risk is the expansion?

After building the spreadsheet and doing the analysis, you can make comparisons and interpret the results. Regardless of the economic outlook or inflation, all four scenarios indicate that expanding the business should provide greater Net Income After Taxes and End-of-year Cash on Hand (in 2017, after two years) than not expanding. So, the DSS model not only provides a quantitative basis for expanding, it provides an analysis that you can present to prospective lenders.

What decision would you make about expansion if you looked only at the 2016 forecast? You could go back to the original spreadsheet and look at the figures for 2016 in the Summary of Key Results section, or you could go to Scenario Manager and create a new summary, specifying the 2016 cells C12 through C15. See Figure C-28.

	A	B	C	D	E	F	G	H
1	Tutorial Exercise--Collegetown Thrift Shop							
2								
3	Constants	2015	2016	2017				
4	Tax Rate	NA	33%	35%				
5	Loan Amount for Store Expansion	NA	$100,000	NA				
6								
7	Inputs	2015	2016	2017				
8	Economic Outlook (R=Recession, B=Boom)	NA	R	NA				
9	Inflation Outlook (H=High, L=Low)	NA	H	NA				
10								
11	Summary of Key Results	2015	2016	2017				
12	Net Income after Taxes (Expansion)	NA	$69,974	$73,660				
13	End-of-year Cash on Hand (Expansion)	NA	$84,974	$158,634				
14	Net Income after Taxes (No Expansion)	NA	$72,601	$69,936				
15	End-of-year Cash on Hand (No Expansion)	NA	$87,601	$157,537				
16								
17	Calculations (Expansion)	2015	2016	2017				
18	Total Sales Dollars	$350,000	$455,000	$591,500				
19	Cost of Goods Sold	$245,000	$337,610	$465,227				
20	Cost of Goods Sold (as a percent of Sales)	70%	74%	79%				

Scenario Summary

Report type
⦿ Scenario summary
◯ Scenario PivotTable report

Result cells:
$C$12:$C$15

[OK]   [Cancel]

Source: Microsoft product screenshots used with permission from Microsoft Corporation

FIGURE C-28 Creating a new Scenario Summary for 2016 instead of 2017

When you click OK, Excel creates a second Scenario Summary (appropriately named Scenario Summary 2), but this time the output values come from 2016, not 2017. After editing and formatting, the 2016 Scenario Summary could look like Figure C-29.

	A	B	C	D	E	F
1						
2	Scenario Summary -- Collegetown Thrift Shop Financial Forecast -- End of Year 2016					
3		Cell Address	Recession-High Inflation	Recession-Low Inflation	Boom-High Inflation	Boom-Low Inflation
5	Changing Cells:					
6	Economic Outlook: R-Recession, B-Boom	$C$8	R	R	B	B
7	Inflation: H-High, L-Low	$C$9	H	L	H	L
8	Result Cells:					
9	Net Income After Taxes-Expansion	$C$12	$69,974	$78,510	$61,316	$68,867
10	End-Of-Year Cash On Hand-Expansion	$C$13	$84,974	$93,510	$76,316	$83,867
11	Net Income After Taxes- No Expansion	$C$14	$72,601	$80,480	$63,526	$70,420
12	End-Of-Year Cash On Hand-No Expansion	$C$15	$87,601	$95,480	$78,526	$85,420

Source: Microsoft product screenshots used with permission from Microsoft Corporation

FIGURE C-29 Scenario Summary for End of Year 2016

As you can see, *not* expanding the business yields slightly better financial results at the end of 2016. As the original Scenario Summary points out, it will take two years for the business expansion to start making more money when compared with not expanding. Note that you could revise the original spreadsheet to copy the columns out to 2018, 2019, and beyond to forecast future income and cash flows. However, note that the accuracy of a forecast becomes less certain as you extend it in time.

Managers must also maintain a healthy skepticism about the validity of their assumptions when formulating a DSS model. Most assumptions about economic outlooks, inflation, and interest rates are really educated guesses. For example, who could have predicted the economic meltdown in 2007? Business DSS models for investments, new product launches, business expansion, or major capital projects commonly look at three possible outcomes: best case, most likely, and worst case. The most likely outcome is based on previous years' data already collected by the firm. The best-case and worst-case outcomes are formulated based on some percentage of performance that falls above or below the most likely scenario. At least these are data-driven forecasts, or what people in the business world call "guessing—with data."

So, how do you reduce risk when making financial decisions based on DSS model results? It helps to formulate the model based on valid data and to use conservative estimates for success. More importantly, collecting pertinent data and tracking the business results *after* deciding to invest or expand can help reduce the risk of failure for the enterprise.

Summary Sheets

When you start working on the Scenario Manager spreadsheet cases later in this book, you will need to know how to manipulate summary sheets and their data. Some of these operations are explained in the following sections.

Rerunning Scenario Manager

The Scenario Summary sheet does not update itself when you change formulas or inputs in the spreadsheet. To get an updated Scenario Summary, you must rerun Scenario Manager, as you did when changing the outputs from 2017 to 2016. Click the Summary button in the Scenario Manager window, verify the results, and then click OK. Another summary sheet is created; Excel numbers them sequentially (Scenario Summary, Scenario Summary 2, etc.), so you do not have to worry about Excel overwriting any of your older summaries. That is why you should rename each summary with a description of the changes.

Deleting Unwanted Scenario Manager Summary Sheets

When working with Scenario Manager, you might produce summary sheets you do not want. To delete an unwanted sheet, move your mouse pointer to the group of sheet tabs at the bottom of the screen and *right*-click the tab of the sheet you want to delete. Click Delete from the menu that appears (see Figure C-30).

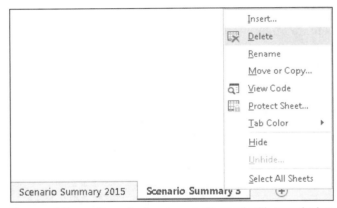

Source: Microsoft product screenshots used with permission from Microsoft Corporation

FIGURE C-30   Deleting unwanted worksheet

Charting Summary Sheet Data

You can easily chart Summary Sheet results using the Charts group in the Insert tab, as discussed in Tutorial F. Figure C-31 shows a 3D clustered column chart prepared from the data in the Scenario Summary for 2017. Charts are useful because they provide a visual comparison of results. As the chart shows, the best economic climate for the thrift shop is a Recession with Low Inflation.

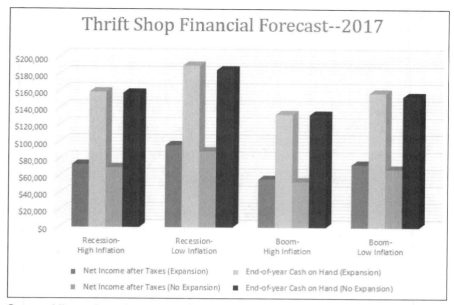

Source: Microsoft product screenshots used with permission from Microsoft Corporation

FIGURE C-31   3D Clustered column chart displaying data from the Summary Sheet

Copying Summary Sheet Data to the Clipboard

As you can with almost everything else in Microsoft Office, you can copy summary sheet data to other Office applications (a Word document or PowerPoint slide, for example) by using the Clipboard. Follow these steps:

1. Select the data range you want to copy.
2. Right-click the mouse and select Copy from the resulting menu.
3. Open the Word document or PowerPoint presentation into which you want to copy.
4. Click your cursor where you want the upper-left corner of the copied data to be displayed.
5. Right-click the mouse and select Paste from the resulting menu. The data should now appear on your document.

PRACTICE EXERCISE—TED AND ALICE'S HOUSE PURCHASE DECISION

Ted and Alice are a young couple who have been living in an apartment for the first two years of their marriage. They would like to buy their first house, but do not know whether they can afford it. Ted works as a carpenter's apprentice, and Alice is a customer service specialist at a local bank. In 2015, Ted's "take home" wages were $24,000 after taxes and deductions, and Alice's take-home salary was $30,000. Ted gets a 2% raise every year, and Alice gets a 3% raise. Their apartment rent is $1,200 per month ($14,400 per year), but the lease is up for renewal and the landlord said he needs to increase the rent for the next lease.

Ted and Alice have been looking at houses and have found one they can buy, but they will need to borrow $200,000 for a mortgage. Their parents are helping them with the down payment and closing costs. After talking to several lenders, Ted and Alice have learned that the state legislature is voting on a first-time home buyers' mortgage bond. If the bill passes, they will be able to get a 30-year fixed mortgage at 3% interest. Otherwise, they will have to pay 6% interest on the mortgage.

Because of the depressed housing market, Ted and Alice are not figuring equity value into their calculations. In addition, although the mortgage interest and real estate taxes will be deductible on their income taxes, these deductions will not be higher than the standard allowable tax deduction, so they are not figuring on any savings there either. Ted and Alice's other living expenses (such as car payments, food, and medical bills), the utilities expenses for either renting or buying, and estimated house maintenance expenses are listed in the Constants section (see Figure C-32).

Ted and Alice's primary concern is their cash on hand at the end of years 2016 and 2017. They are thinking of starting a family, but they know it will be difficult without adequate savings.

Getting Started on the Practice Exercise

If you closed Excel after the first tutorial exercise, start Excel again, then click Blank workbook to begin a new one. If your Excel workbook from the first tutorial is still open, you may find it useful to start a new worksheet in the same workbook. Then you can refer back to the first tutorial when you need to structure or format the spreadsheet; the formatting of both exercises in this tutorial is similar. Set up your new worksheet as explained in the following sections.

You can also download the spreadsheet skeleton if you prefer; it will save you time. To access this skeleton file, select Tutorial C from your data files, and then select **Rent or Buy Skeleton.xlsx**.

Constants Section

Your spreadsheet should have the constants shown in Figure C-32. An explanation of the line items follows the figure.

	A	B	C	D
1	Tutorial Exercise Skeleton--Ted and Alice's House Decision			
2				
3	Constants	2015	2016	2017
4	Non-Housing Living Expenses (Cars, Food, Medical, etc)	NA	$36,000	$39,000
5	Mortgage Amount for Home Purchase	NA	$200,000	NA
6	Real Estate Taxes and Insurance on Home	NA	$3,000	$3,150
7	Utilities Expense (Heat & Electric)--Apartment	NA	$2,000	$2,200
8	Utilities Expense(Heat, Electric, Water, Trash)--House	NA	$2,500	$2,600
9	House Repair and Maintenance Expenses	NA	$1,200	$1,400

Source: Microsoft product screenshots used with permission from Microsoft Corporation

FIGURE C-32   Constants section

- Non-Housing Living Expenses—This value represents Ted and Alice's estimate of all their other living expenses for 2016 and 2017.
- Mortgage Amount for Home Purchase
- Real Estate Taxes and Insurance on Home—A lender has given Ted and Alice estimates for these values; they are usually paid monthly with the house mortgage payment. The money is placed in an escrow account and then paid by the mortgage company to the state or county and insurance company.
- Utilities Expense—Apartment—This value is Ted and Alice's estimate for 2016 and 2017 based on their 2015 bills.
- Utilities Expense—House—Currently the apartment rent includes fees for water, sewer, and trash disposal. If they get a house, Ted and Alice expect the utilities to be higher.
- House Repair and Maintenance Expenses—In an apartment, the landlord is responsible for repair and maintenance. Ted and Alice will have to budget for repair and maintenance on the house.

Inputs Section

Your spreadsheet should have the inputs shown in Figure C-33. An explanation of line items follows the figure.

	A	B	C	D
11	Inputs	2015	2016	2017
12	Rental Occupancy (H=High, L=Low)	NA		NA
13	First Time Buyer Bond Loans Available (Y=Yes, N=No)	NA		NA

Source: Microsoft product screenshots used with permission from Microsoft Corporation

FIGURE C-33   Inputs section

- Rental Occupancy (H=High, L=Low)—When the housing market is depressed (in other words, people are not buying homes), rental housing occupancy percentages are high, which allows landlords to charge higher rents when leases are renewed. Ted and Alice think their rent will increase in 2016. The amount of the increase depends on the Rental Occupancy. If the

occupancy is high, Ted and Alice expect to see a 10% increase in rent in both 2016 and 2017. If occupancy is low, they only expect a 3% increase for each year.
- First Time Buyer Bond Loans Available (Y=Yes, N=No)—As described earlier, when housing markets are depressed, local governments will frequently pass a bond bill to provide low-interest mortgage money to first-time home buyers. If the bond loans are available, Ted and Alice can obtain a 30-year fixed mortgage at only 3%, which is half the interest rate they would otherwise pay for a conventional mortgage.

Summary of Key Results Section

Figure C-34 shows what key results Ted and Alice are looking for. They want to know their End-of-year Cash on Hand for both 2016 and 2017 if they decide to stay in the apartment and if they decide to purchase the house.

	A	B	C	D
15	**Summary of Key Results**	**2015**	**2016**	**2017**
16	End-of-year Cash on Hand (Rent)	NA		
17	End-of-year Cash on Hand (Buy)	NA		

Source: Microsoft product screenshots used with permission from Microsoft Corporation

FIGURE C-34 Summary of Key Results section

These results are copied from the End-of-year Cash on Hand sections of the Income and Cash Flow Statements sections (for both renting and buying).

Calculations Section

Your spreadsheet will need formulas to calculate the apartment rent, house payments, and interest rate for the mortgage (see Figure C-35). You will use the rent and house payments later in the Income and Cash Flow Statements for both renting and buying.

	A	B	C	D
19	**Calculations**	**2015**	**2016**	**2017**
20	Apartment Rent	$14,400		
21	House Payments	NA		
22	Interest Rate for House Mortgage		NA	NA

Source: Microsoft product screenshots used with permission from Microsoft Corporation

FIGURE C-35 Calculations section

- Apartment Rent—The 2015 amount is given. Use IF formulas to increase the rent by 10% if occupancy rates are high, or by 3% if occupancy rates are low.
- House Payments—This value is the total of the 12 monthly payments made on the mortgage. An important point to note is that house mortgage interest is always compounded *monthly*, not annually, as in the thrift shop tutorial. To properly calculate the house payments for the year, you divide the annual interest rate by 12 to determine the monthly interest. You also have to multiply a 30-year mortgage by 12 to get 360 payments, and then multiply the PMT formula by 12 to get the total amount for your annual house payments. Also, you will precede the PMT function with a negative sign to make the payment amount a positive number. Your formula should look like the following:
  =–PMT(B22/12,360,C5)*12
- Interest Rate for House Mortgage—Use the IF formula to enter a 3% interest rate if the bond money is available, and a 6% interest rate if no bond money is available.

Income and Cash Flow Statements Sections

As with the thrift shop tutorial, you want to see the Income and Cash Flow Statements for two scenarios—in this case, for continuing to rent and for purchasing a house. Each section begins with cash on hand at the end of 2015. As you can see in Figure C-36, Ted and Alice have only $4,000 in their savings.

	A	B	C	D
24	**Income and Cash Flow Statement (Continue to Rent)**	**2015**	**2016**	**2017**
25	Beginning-of-year Cash on Hand	NA		
26	Ted's Take Home Wages	$24,000		
27	Alice's Take Home Salary	$30,000		
28	Total Take Home Income	$54,000		
29	*Apartment Rent*	NA		
30	Utilities (Apartment)	NA		
31	Non-Housing Living Expenses	NA		
32	Total Expenses	NA		
33	End-of-year Cash on Hand	$4,000		
34				
35	**Income and Cash Flow Statement (Purchase House)**	**2015**	**2016**	**2017**
36	Beginning-of-year Cash on Hand	NA		
37	Ted's Take Home Wages	$24,000		
38	Alice's Take Home Salary	$30,000		
39	Total Take Home Income	$54,000		
40	*House Payments*	NA		
41	*Real Estate Taxes and Insurance*	NA		
42	Utilities (House)	NA		
43	*House Repair and Maintenance Expense*	NA		
44	Non-Housing Living Expenses	NA		
45	Total Expenses	NA		
46	End-of-year Cash on Hand	$4,000		

Source: Microsoft product screenshots used with permission from Microsoft Corporation

FIGURE C-36   Income and Cash Flow Statements sections (for both rent and purchase)

- Beginning-of-year Cash on Hand—This value is the End-of-year Cash on Hand from the previous year.
- Ted's Take Home Wages—This value is given for 2015. To get values for 2016 and 2017, increase Ted's wages by 2% each year.
- Alice's Take Home Salary—This value is given for 2015. To get values for 2016 and 2017, increase Alice's salary by 3% each year.
- Total Take Home Income—The sum of Ted and Alice's pay.
- Apartment Rent—The rent is copied from the Calculations section.
- House Payments—The house payments are also copied from the Calculations section.
- Real Estate Taxes and Insurance, Utilities (Apartment or House), House Repair and Maintenance Expense, and Non-Housing Living Expenses—These values all are copied from the Constants section.
- Total Expenses—This value is the sum of all the expenses listed above. Note that the house payment is now a positive number, so you can sum it normally with the other expenses.
- End-of-year Cash on Hand—This value is the Beginning-of-year Cash on Hand plus the Total Take Home Income minus the Total Expenses.

Scenario Manager Analysis

When you have completed the spreadsheet, set up Scenario Manager and create a Scenario Summary sheet. Ted and Alice want to look at their End-of-year Cash on Hand in 2017 for renting or buying under the following four scenarios:

- High occupancy and bond money available
- High occupancy and no bond money available
- Low occupancy and bond money available
- Low occupancy and no bond money available

If you have done your spreadsheet and Scenario Manager correctly, you should get the results shown in Figure C-37.

	A	B	C	D	E	F
1	Scenario Summary--Ted & Alice's House Purchase Decision--2017					
2			Hi Occ- Bond $	Hi Occ- No Bond $	Lo Occ- Bond $	Lo Occ- No Bond $
4	Changing Cells:					
5	Rental Occupancy (H-High, L-Low)	$C$12	H	H	L	L
6	Bond Mortgage Available (Y or N)	$C$13	Y	N	Y	N
7	Result Cells:					
8	End of Year Cash on Hand (Rent)	$D$16	$3,713	$3,713	$6,868	$6,868
9	End of Year Cash on Hand (Buy)	$D$17	$7,090	($1,452)	$7,090	($1,452)

Source: Microsoft product screenshots used with permission from Microsoft Corporation

FIGURE C-37   Scenario Summary results

Interpreting the Results

Based on the Scenario Summary results, what should Ted and Alice do? At first glance, it looks like the safe decision is to stay in the apartment. Actually, their decision hinges on whether they can get the lower-interest mortgage from the first-time buyers' bond issue. If they can, and if occupancy levels in apartments stay high, purchasing a house will give them about $3,300 more in savings at the end of 2017 than if they continued renting. Some other intangible factors are that home owners do not need permission to have pets, detached houses are quieter than apartments, and homes usually have a yard for pets and children to play in. Also, for the purposes of this exercise, you did not consider the tax benefits of home ownership. Depending on the amount of mortgage interest and real estate taxes Ted and Alice have to pay, they may be able to itemize their deductions and pay less income tax. If the income tax savings are more than $1,500, they can purchase the house even at the higher interest rate. In any case, because you did the DSS model for them, Ted and Alice now have a quantitative basis to help them make a good decision.

Visual Impact: Charting the Results

Charts and graphs often add visual impact to a Scenario Summary. Using the data from the Scenario Summary output table, try to create a chart similar to the one in Figure C-38 to illustrate the financial impact of each outcome.

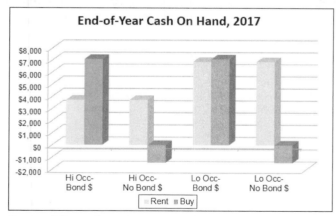

Source: Microsoft product screenshots used with permission from Microsoft Corporation

FIGURE C-38   A 3D clustered column chart created from Scenario Summary data

Printing and Submitting Your Work

Ask your instructor which worksheets need to be printed for submission. Make sure your printouts of the spreadsheet, the Scenario Manager Summary table, and the graph (if you created one) fit on one printed page apiece. Click the File tab on the Ribbon, click the Print button, and then click Page Setup at the bottom of the Print Navigation pane. When the Page Setup window opens, click the Page tab if it is not already open, then click the Fit to radio button and click 1 page wide by 1 page tall. Your spreadsheet, table, and graph will be fitted to print on one page apiece.

REVIEW OF EXCEL BASICS

This section reviews some basic operations in Excel and provides some tips for good work practices. Then you will work through some cash flow calculations. Working through this section will help you complete the spreadsheet cases in the following chapters.

Save Your Work Often—and in More Than One Place

To guard against data loss in case of power outages, computer crashes, and hard drive failure, it is always a good idea to save your work to a separate storage device. Copying a file into two separate folders on the same hard disk is *not* an adequate safeguard. If you are working on your college's computer network and you have been assigned network storage, the network storage is usually "mirrored"; in other words, it has duplicate drives recording data to prevent data loss if the system goes down. However, most laptops and home computers lack this feature. An excellent way to protect your work from accidental deletion is to purchase a USB "thumb" drive and copy all of your files to it.

When you save your Excel files, Windows will usually store them in the My Documents folder unless you specify the storage location. Instead of just clicking the Save icon, click the File group in the Ribbon, click Save As in the left column menu, and then click the Browse button. A window will appear with icons on the left side, as shown in Figure C-39. If you have previously saved your file to a particular location, it will appear in the Save in text box at the top of the window. To save the file in the same location, click Save. If your work is stored elsewhere, you can find the location using the icons on the left side of the window. If you are saving to a USB thumb drive, it will appear as a storage device when you click the Computer icon. Click the folder where you want to save your file.

NOTE

If you are trying an operation that might damage your spreadsheet and you do not want to use the Undo command, you can use the Save As command, and then add a number or letter to the filename to save an additional copy to "play with." Your original work will be preserved.

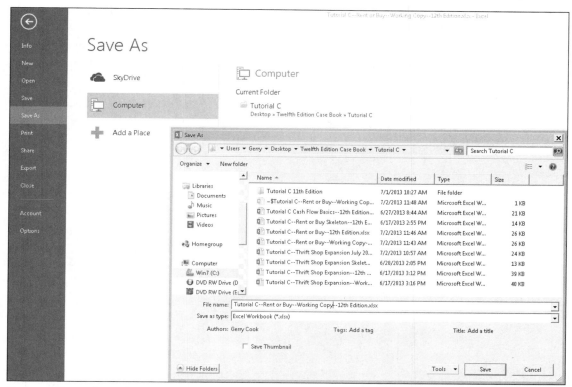

Source: Microsoft product screenshots used with permission from Microsoft Corporation

FIGURE C-39   The Save As window in Excel

Basic Operations

To begin, you will review the following topics: formatting cells, displaying the spreadsheet cell formulas, circular reference errors, using the AND and OR logical operators in IF statements, and using nested IF statements to produce more than two outcomes.

Formatting Cells

Cell Alignment

Headings for columns are usually centered, while numbers in cells are usually aligned to the right. To set the alignment of cell data:

1. Highlight the cell or cell range to format.
2. Select the Home tab.
3. In the Alignment group, click the button representing the horizontal alignment you want for the cell (Align Left, Center, or Align Right).
4. Also in the Alignment group, above the horizontal alignment buttons, click the vertical alignment you want (Top Align, Middle Align, or Bottom Align). Middle Align is the most common vertical alignment for cells.

Cell Borders

Bottom borders are common for headings, and accountants include borders and double borders to indicate subtotals and grand totals on spreadsheets. Sometimes it is also useful to put a "box" border around a table of values or a section of a spreadsheet. To create borders:

1. Highlight the cell or cell range that needs a border.
2. Select the Home tab.

3. In the Font group, click the drop-down arrow of the Border icon. A menu of border selections appears (see Figure C-40).
4. Choose the desired border for the cell or group of cells. Note that All Borders creates a box border around each cell, while Outside Borders draws a box around a group of cells.

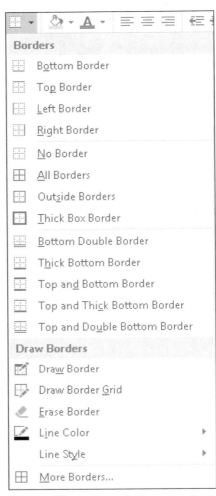

Source: Microsoft product screenshots used with permission from Microsoft Corporation

FIGURE C-40   Selections in the Borders menu

Number Formats

For financial numbers, you usually use the Currency format. (Do not use the Accounting format, as it places the $ sign to the far left side of the cell). To apply the appropriate Currency format:

1. Highlight the cell or cell range to be formatted.
2. Select the Home tab.
3. In the Number group, select Currency in the Number Format drop-down list.
4. To set the desired number of decimal places, click the Increase Decimal or Decrease Decimal button in the bottom-right corner of the group (see Figure C-41).

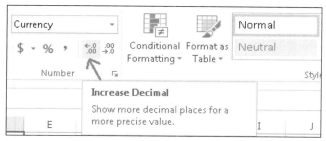

Source: Microsoft product screenshots used with permission from Microsoft Corporation

FIGURE C-41   Increase Decimal and Decrease Decimal buttons

If you do not know what a button does in Office, hover your mouse pointer over the button to see a description.

Format "Painting"

If you want to copy *all* the format properties of a certain cell to other cells, use the Format Painter. First, select the cell whose format you want to copy. Then click the Format Painter button (the paintbrush icon) in the Clipboard group under the Home tab (see Figure C-42). When you click the button, the mouse pointer turns into a paintbrush. Click the cell you want to reformat. To format multiple cells, select the cell whose format you want to copy, and then click *twice* on the Format Painter button. The mouse cursor will become a paintbrush, and the paint function will stay on so you can reformat as many cells as you want. To turn off the Format Painter, click its button again or press the Esc key.

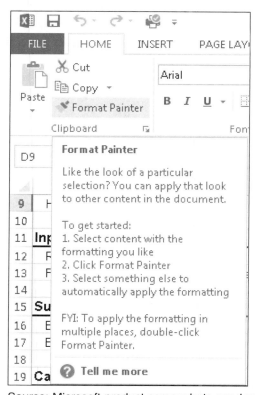

Source: Microsoft product screenshots used with permission from Microsoft Corporation

FIGURE C-42   The Format Painter button

Showing the Excel Formulas in the Cells

Sometimes your instructor might want you to display or print the formulas in the spreadsheet cells. If you want the spreadsheet cells to display the actual cell formulas, follow these steps:

1. While holding down the Ctrl key, press the key in the upper-left corner of the keyboard that contains the back quote (`) and tilde (~). The spreadsheet will display the formulas in the cells. The columns may also become quite wide—if so, do not resize them.

2. The Ctrl+`~ key combination is a toggle; to restore your spreadsheet to the normal cell contents, press Ctrl+`~ again.

Understanding Circular Reference Errors

When entering formulas, you might make the mistake of referring to the cell in which you are entering the formula as part of the formula, even though it should only display the output of that formula. Referring a cell back to itself in a formula is called a *circular reference*. For example, suppose that in cell B2 of a worksheet, you enter =B2-B1. A terrible but apt analogy for a circular reference is a cannibal trying to eat himself! Excel 2010 would inform you when you tried to enter a circular reference into a formula, and would warn you if you tried to open an existing spreadsheet that has one or more circular references.

Excel 2013 has changed its treatment of circular reference errors. It allows you to enter them without warning, displaying a 0 in the cell, unless you do the following:

1. Click the File tab.
2. Click Options in the left column menu.
3. Click Formulas in the left column of the Excel Options window.
4. Uncheck the "Enable iterative calculation" check box (see Figure C-43), and then click OK. Now, when you enter a circular reference in the spreadsheet, you receive a warning message (see Figure C-44).

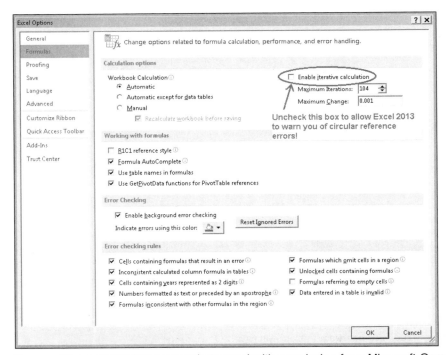

Source: Microsoft product screenshots used with permission from Microsoft Corporation

FIGURE C-43   Setting formula options in Excel

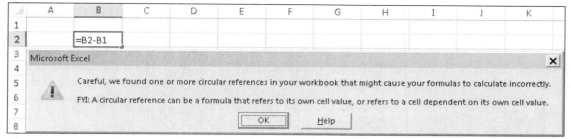

Source: Microsoft product screenshots used with permission from Microsoft Corporation

FIGURE C-44   Excel circular reference warning

Using AND and OR Functions in IF Statements

Recall that the IF function has the following syntax:

IF(test condition, result if test is True, result if test is False)

The test conditions in the previous example IF statements tested only one cell's value, but a test condition can test more than one value of a cell.

For example, look at the thrift shop tutorial again. The Total Sales Dollars for 2016 depended on the economic outlook (recession or boom). The original IF statement was =IF($C$8="R",B18*1.3,B18*1.15), as shown in Figure C-45. This function increased the baseline 2015 Total Sales Dollars by 30% if there was continued recession ("R" entered in cell C8), but only increased the total by 15% if there was a boom.

C18	⌄	:	×	✓	f_x	=IF($C$8="R",B18*1.3,B18*1.15)		

	A	B	C	D
1	**Tutorial Exercise--Collegetown Thrift Shop**			
2				
3	**Constants**	**2015**	**2016**	**2017**
4	Tax Rate	NA	33%	35%
5	Loan Amount for Store Expansion	NA	$100,000	NA
6				
7	**Inputs**	**2015**	**2016**	**2017**
8	Economic Outlook (R=Recession, B=Boom)	NA	R	NA
9	Inflation Outlook (H=High, L=Low)	NA	H	NA
10				
11	**Summary of Key Results**	**2015**	**2016**	**2017**
12	Net Income after Taxes (Expansion)	NA	$69,974	$73,660
13	End-of-year Cash on Hand (Expansion)	NA	$84,974	$158,634
14	Net Income after Taxes (No Expansion)	NA	$72,601	$69,936
15	End-of-year Cash on Hand (No Expansion)	NA	$87,601	$157,537
16				
17	**Calculations (Expansion)**	**2015**	**2016**	**2017**
18	Total Sales Dollars	$350,000	$455,000	$591,500
19	Cost of Goods Sold	$245,000	$337,610	$465,227
20	Cost of Goods Sold (as a percent of Sales)	70%	74%	79%
21	Interest Rate for Business Loan	5%	NA	NA

Source: Microsoft product screenshots used with permission from Microsoft Corporation

FIGURE C-45   The original IF statement used to calculate the Total Sales Dollars for 2016

To take the IF argument one step further, assume that the Total Sales Dollars depended not only on the Economic Outlook, but on the Inflation Outlook (High or Low). Suppose there are two possibilities:

- Possibility 1: If the economic outlook is for a Recession and the inflation outlook is High, the Total Sales Dollars for 2016 will be 30% higher than in 2015.
- Possibility 2: For the other three cases (Recession and Low Inflation, Boom and High Inflation, and Boom and Low Inflation), assume that the 2016 Total Sales Dollars will only be 15% higher than in 2015.

The first possibility requires two conditions to be true at the same time: C8="R" and C9="H". You can include an AND( ) function inside the IF statement to reflect the additional condition as follows:

=IF(AND($C$8="R",$C$9="H"), B18*130%,B18*115%)

When the test argument uses the AND() function, conditions "R *and* "H" both must be present at the same time for the statement to use the true result (thus multiplying last year's sales by 130%). Any of the other three outcome combinations will cause the statement to use the false result (thus multiplying last year's sales by 115%).

You can also use an OR() function in an IF statement. For example, assume that instead of both conditions (Recession and High Inflation) having to be present, only one of the two conditions needs to be present for sales to increase by 30%. In this case, you use the OR() function in the test argument as follows:

=IF(OR($C$8="R",$C$9="H"), B18*130%,B18*115%)

In this case, if *either* of the two conditions (C8="R" or C9="H"), is true, the function will return the true argument, multiplying the 2015 sales by 130%. If *neither* of the two conditions is true, then the function will return the false argument, multiplying the 2015 sales by 115% instead.

Using IF Statements Inside IF Statements (Also Called "Nesting IFs")

By now you should be familiar with IF statements, but here is a quick review of the syntax:

=IF(test condition, result if test is True, result if test is False)

In the preceding examples, only two courses of action were possible for each of the inputs: Recession or Boom, High Inflation or Low Inflation, Rental Occupancy High or Low, Bond Money Available or No Bond Money Available. The tutorial used only two possible outcomes to keep them simple.

However, in the business world, decision support models are frequently based on three or more possible outcomes. For capital projects and new product launches, you will frequently project financial outcomes based on three possible scenarios: Most Likely, Worst Case, and Best Case. You can modify the IF statement by placing another IF statement inside the result argument if the first test is false, creating the ability to launch two more alternatives from the second IF statement. This is called "nesting" your IF statements.

Try a simple nested IF statement: In your thrift shop example, assume that three economic outlooks are possible: Recession (R), Boom (B), or Stable (S). As before, the 2016 Total Sales Dollars (cell C18) will be the 2015 Total Sales Dollars increased by some fixed percentage. In a Recession, sales will increase by 30%, in a Boom they will increase by 15%, and for a Stable Economic Outlook, sales will increase by 22%, which is roughly midway between the other two percentages. You can "nest" the IF statement in cell C18 to reflect the third outcome as follows:

=IF($C$8= "R",B18*130%,IF($C$8="B",B18*115%,B18*122%))

Note the added IF statement inside the False value argument. You can break down this statement:

- If the value in cell C8 is "R", multiply the value in cell B18 by 130%, and enter the result in cell C18.
- If the value in cell C8 is not "R", check whether the value in cell C8 is "B". If it is "B", multiply the value in cell B18 by 115%, and enter the result in cell C18.
- If the value in cell C8 is not "R" and it is not "B", then it must be "S", so multiply the value in cell B18 by 122%, and enter the result in cell C18.

If you have four or more alternatives, you can keep nesting IF statements inside the false argument for the outer IF statements. (Excel 2007 and later versions have a limit of 64 levels of nesting in the IF function, which should take care of every conceivable situation.)

NOTE

The "embedded IFs" in a nested IF statement are not preceded by an equals sign. Only the first IF gets the equals sign.

Cash Flow Calculations: Borrowing and Repayments

The Scenario Manager cases that follow in this book require accounting for money that the fictional company will have to borrow or repay. This money is not like the long-term loan that the Collegetown Thrift Shop is considering for its expansion. Instead, this money is short-term borrowing that companies use to pay current obligations, such as purchasing inventory or raw materials. Such short-term borrowing is called a line of credit, and is extended to businesses by banks, much like consumers have credit cards. We want to focus on how to do short-term borrowing and repayment calculations, like those you will see in your cases.

To work through cash flow calculations, you must make two assumptions about a company's borrowing and repayment of short-term debt. First, you assume that the company has a desired *minimum* cash level at the end of a fiscal year (which is also its cash level at the start of the next fiscal year), to ensure that the company can cover short-term expenses and purchases. Second, assume the bank that serves the company will provide short-term loans (a line of credit) to make up the shortfall if the end-of-year cash falls below the desired minimum level.

NCP stands for Net Cash Position, which equals beginning-of-year cash plus net income after taxes for the year. NCP represents the available cash at the end of the year, *before* any borrowing or repayment. For the three examples shown in Figure C-46, set up a simple spreadsheet in Excel and determine how much the company needs to borrow to reach its minimum year-end cash level. Use the IF function to enter 0 under Amount to Borrow if the company does not need to borrow any money.

	A	B	C	D
	Example	NCP	Minimum Cash Required	Amount To Borrow
1	Example	NCP	Minimum Cash Required	Amount To Borrow
2	1	$ 25,000	$ 10,000	?
3	2	$ 9,000	$ 10,000	?
4	3	$ (12,000)	$ 10,000	?

Source: Microsoft product screenshots used with permission from Microsoft Corporation

FIGURE C-46 Examples of borrowing

You can also assume that the company will use some of its cash on hand at the end of the year to pay off as much of its outstanding debt as possible without going below its minimum cash on hand required. The "excess" cash is the company's NCP *less* the minimum cash on hand required—any cash above the minimum is available to repay any debt. In the examples shown in Figure C-47, compute the excess cash and then compute the amount to repay. In addition, compute the ending cash on hand after the debt repayment.

	A	B	C	D	E	F
	Example	NCP	Minimum Cash Required	Beginning of Year Debt	Repay?	Ending Cash
1	Example	NCP	Minimum Cash Required	Beginning of Year Debt	Repay?	Ending Cash
2	1	$ 12,000	$ 10,000	$ 5,000	?	?
3	2	$ 13,000	$ 10,000	$ 8,000	?	?
4	3	$ 20,000	$ 10,000	$ -	?	?
5	4	$ 60,000	$ 10,000	$ 40,000	?	?
6	5	$ (20,000)	$ 10,000	$ 10,000	?	?

Source: Microsoft product screenshots used with permission from Microsoft Corporation

FIGURE C-47 Examples of debt repayment

In the Scenario Manager cases of the following chapters, your spreadsheets may need two bank financing sections beneath the Income and Cash Flow Statements sections. You will build the first section to calculate any needed borrowing or repayment at year's end to compute year-end cash on hand. The second section will calculate the amount of debt owed at the end of the year after any borrowing or repayment.

Return to the Collegetown Thrift Shop tutorial and assume that it includes a line of credit at a local bank for short-term cash management. The first new section extends the end-of-year cash calculation, which was shown for the thrift shop in Figure C-19. Figure C-48 shows the structure of the new section highlighted in boldface.

	A	B	C	D
29	**Income and Cash Flow Statements (Expansion)**	**2015**	**2016**	**2017**
30	Beginning-of-year Cash on Hand	NA	$15,000	$0
31	Sales (Revenue)	NA	$455,000	$591,500
32	Cost of Goods Sold	NA	$337,610	$465,227
33	*Business Loan Payment*	NA	($12,950)	($12,950)
34	Income before Taxes	NA	$104,440	$113,323
35	Income Tax Expense	NA	$34,465	$39,663
36	Net Income after Taxes	NA	$69,974	$73,660
37	**Net Cash Position NCP=** **Beginning-of-year Cash on Hand** **plus Net Income after Taxes**	NA	$84,974	$73,660
38	**Line of credit borrowing from bank**	NA		
39	**Line of credit repayments to bank**	NA		
40	**End-of Year Cash on Hand**	$15,000		

Source: Microsoft product screenshots used with permission from Microsoft Corporation

FIGURE C-48 Calculation section for End-of-Year Cash on Hand with borrowing and repayments added

The heading in cell A36 was originally End-of-year Cash on Hand in Figure C-19, but you will add line-of-credit borrowing and repayment to the end-of-year totals. You must add the line-of-credit borrowing from the bank to the NCP and subtract the line-of-credit repayments to the bank from the NCP to obtain the End-of-Year Cash on Hand.

The second new section you add will compute the End-of-year debt owed. This section is called Debt Owed, as shown in Figure C-49.

	A	B	C	D
42	**Debt Owed**	**2015**	**2016**	**2017**
43	**Beginning-of-year debt owed**	NA		
44	**Borrowing from bank line of credit**	NA		
45	**Repayment to bank line of credit**	NA		
46	**End-of-year debt owed**	$47,000		

Source: Microsoft product screenshots used with permission from Microsoft Corporation

FIGURE C-49 Debt Owed section

As you can see, the thrift shop currently owes $47,000 on its line of credit at the end of 2015. The End-of-year debt owed equals the Beginning-of-year debt owed plus any new borrowing from the bank's line of credit, minus any repayment to the bank's line of credit. Therefore, the formula in cell C46 would be:

=C43+C44-C45

Assume that the amounts for borrowing and repayment (cells C44 and C45) were calculated in the first new section (for the year 2016, the amounts would be in cells C38 and C39), and then copied into the second section. The formula for cell C44 would be =C38, and for cell C45 would be =C39. The formula for cell C43, Beginning-of-year debt owed in 2016, would simply be the End-of-year debt owed in 2015, or =B46.

Now that you have added the spreadsheet entries for borrowing and repayment, consider the logic for the borrowing and repayment formulas.

Calculation of Borrowing from the Bank Line of Credit

When using logical statements, it is sometimes easier to state the logic in plain language and then turn it into an Excel formula. For borrowing, the logic in plain language is:

If (cash on hand before financing transactions is greater than the minimum cash required,

then borrowing is not needed; else,

borrow enough to get to the minimum)

You can restate this logic as the following:

If (NCP is greater than minimum cash required,

then borrowing from bank=0;

else, borrow enough to get to the minimum)

You have not added minimum cash at the end of the year as a requirement, but you could add it to the Constants section at the top of the spreadsheet (in this case the new entry would be cell C6). Assume that you want $50,000 as the minimum cash on hand at the end of both 2016 and 2017. Assuming that the NCP is shown in cell C37, you could restate the formula for borrowing (cell C38) as the following:

IF(NCP>Minimum Cash, 0; otherwise, borrow enough to get to the minimum cash)

You have cell addresses for NCP (cell C37) and for Minimum Cash (cell $C$6). To develop the formula for cell C38, substitute the cell address for the test argument; the true argument is simply zero (0), and the false argument is the minimum cash minus the current NCP. The formula stated in Excel for cell C38 would be:

=IF(C37>=$C$6, 0, $C$6-C37)

Calculation of Repayment to the Bank Line of Credit

Simplify the statements first in plain language:

IF(beginning of year debt=0, repay 0 because nothing is owed, but

IF(NCP is less than the minimum, repay 0, because you must borrow, but

IF(extra cash equals or exceeds the debt, repay the whole debt,

ELSE (to stay above the minimum cash, repay the extra cash above the minimum)

Look at the following formula. If you assume that the repayment amount will be in cell C39, the beginning-of-year debt is in cell C43, and the minimum cash target is still in cell $C$6, the repayment formula for cell C39 with the nested IFs should look like the following:

=IF(C43=0,0,IF(C37<=$C$6,0,IF((C37-$C$6)>=C43,C43,C37-$C$6)))

The new sections of the thrift shop spreadsheet would look like those in Figure C-50.

C39	fx	=IF(C43=0,0,IF(C37<=$C$6,0,IF((C37-$C$6)>=C43,C43,C37-$C$6)

	A	B	C	D
29	Income and Cash Flow Statements (Expansion)	2015	2016	2017
30	Beginning-of-year Cash on Hand	NA	$15,000	$50,000
31	Sales (Revenue)	NA	$455,000	$591,500
32	Cost of Goods Sold	NA	$337,610	$465,227
33	Business Loan Payment	NA	($12,950)	($12,950)
34	Income before Taxes	NA	$104,440	$113,323
35	Income Tax Expense	NA	$34,465	$39,663
36	Net Income after Taxes	NA	$69,974	$73,660
37	Net Cash Position NCP Beginning-of-year Cash on Hand plus Net Income after Taxes	NA	$84,974	$123,660
38	Line of credit borrowing from bank	NA	$0	$0
39	Line of credit repayments to bank	NA	$34,974	$12,026
40	End-of Year Cash on Hand	$15,000	$50,000	$111,634
41				
42	Debt Owed	2015	2016	2017
43	Beginning-of-year debt owed	NA	$47,000	$12,026
44	Borrowing from bank line of credit	NA	$0	$0
45	Repayment to bank line of credit	NA	$34,974	$12,026
46	End-of-year debt owed	$47,000	$12,026	$0

Source: Microsoft product screenshots used with permission from Microsoft Corporation

FIGURE C-50 Thrift shop spreadsheet with line-of-credit borrowing, repayments, and Debt Owed added

Answers to the Questions about Borrowing and Repayment

Figures C-51 and C-52 display solutions for the borrowing and repayment calculations.

	A	B	C	D
1	Example	NCP	Minimum Cash Required	Amount To Borrow
2	1	$ 25,000	$ 10,000	$ -
3	2	$ 9,000	$ 10,000	$ 1,000
4	3	$ (12,000)	$ 10,000	$ 22,000

Source: Microsoft product screenshots used with permission from Microsoft Corporation

FIGURE C-51   Answers to examples of borrowing

In Figure C-51, you can see that the formula in cell D2 for the amount to borrow is =IF(B2>=C2,0,C2-B2).

E2 f_x =IF(D2=0,0,IF(B2<C2,0,IF((B2-C2)>=D2,D2,B2-C2)))

	A	B	C	D	E	F
1	Example	NCP	Minimum Cash Required	Beginning of Year Debt	Repay?	Ending Cash
2	1	$ 12,000	$ 10,000	$ 5,000	$ 2,000	$ 10,000
3	2	$ 13,000	$ 10,000	$ 8,000	$ 3,000	$ 10,000
4	3	$ 20,000	$ 10,000	$ -	$ -	$ 20,000
5	4	$ 60,000	$ 10,000	$ 40,000	$ 40,000	$ 20,000
6	5	$ (20,000)	$ 10,000	$ 10,000	$ -	$ (20,000)

Source: Microsoft product screenshots used with permission from Microsoft Corporation

FIGURE C-52   Answers to examples of repayment

Note that the repayment formula for cell E2 is shown in Figure C-52.
    Note the following points about the repayment calculations shown in Figure C-52.

- In Example 1, only $2,000 is available for debt repayment ($12,000 – $10,000) to avoid dropping below the Minimum Cash Required.
- In Example 2, only $3,000 is available for debt repayment.
- In Example 3, the Beginning-of-Year Debt was zero, so the Ending Cash is the same as the Net Cash Position.
- In Example 4, there was enough cash to repay the entire $40,000 debt, leaving $20,000 in Ending Cash.
- In Example 5, the company has cash problems—it cannot repay any of the Beginning-of-Year Debt of $10,000, and it will have to borrow an additional $30,000 to reach the Minimum Cash Required target of $10,000.

You should now have all the basic tools you need to tackle Scenario Manager in Cases 6 and 7. Good luck!

TRIPOD MARKETING INC.

Decision Support Using Microsoft Excel

PREVIEW

As a new summer intern, you have been assigned to create a forecasting tool to estimate funds required by Tripod Marketing in 2015. In this case, you will use Microsoft Excel to provide your management team a look into the future.

PREPARATION

- Review spreadsheet concepts discussed in class and in your textbook.
- Complete any exercises that your instructor assigns.
- Complete any part of Tutorial C that your instructor assigns. You may need to review the use of If statements and the section called "Cash Flow Calculations: Borrowing and Repayments."
- Review file-saving procedures for Windows programs.
- Refer to Tutorials E and F as necessary.

BACKGROUND

You did it! You were selected for the coveted summer internship at Tripod Marketing Inc. (TMI). On your first day, you are welcomed to the company's campus by CEO Robert Fleckburg and Chief Marketing Officer Jon Flookstein, who are excited to have you on board. They explain that your main task for the summer will be to create a forecast modeling tool with Excel that allows them to make budget requisitions and determine how different scenarios would affect the department's budget. Your work will help the Marketing department forecast its 2015 budget needs so it can request working capital from the Finance department.

TMI is a world leader in direct-mail campaign design and execution. Direct-mail marketing is a method of sending communications to potential customers via the United States Postal Service (USPS). TMI offers three different types of envelopes that vary in size and cost to better suit its customers' needs.

- The Jumbo envelope is the most expensive because of its size and composition. This 8- by 10-inch envelope is made of a thin, cardboard-like material, much like a priority envelope used by the USPS. This option offers clients a more official look than smaller paper envelopes.
- The 6- by 9-inch envelope is the mid-sized option. It stands out from regular envelopes. The 6×9 envelope is also the middle option in terms of price.
- The #10 envelope is the most commonly used envelope for invoices, notices, or other simple documents. At 4 by 9 inches, it is the most economical alternative of the three envelope options.

Jon says your work will be important for three direct-mail campaigns planned during 2015:

- ABC Corp. will use Jumbo and 6×9 envelopes for one campaign.
- Delaware Inc. will use 6×9 and #10 envelopes for another campaign.
- The Blue Hens Fan Club will use the #10 envelope for a third campaign.

Jon also provides several important numbers that you will need for your forecasts:

- The amount of Jumbo envelopes in stock at the beginning of the year
- The amount of 6×9 envelopes in stock at the beginning of the year
- The amount of #10 envelopes in stock at the beginning of the year
- The handling costs of Jumbo envelopes

- The handling costs of 6 × 9 envelopes
- The handling costs of #10 envelopes
- A discount postage rate for high-volume campaigns
- The first-class postage rate for presorted mail
- The first-class postage rate for unsorted mail
- A potential postage credit from the USPS

Assignment 1 contains information you need to write the formulas for the Calculations section, including formulas for inventory, campaign costs, and mail volumes. Jon has been working with the USPS to obtain a postage credit of $25,000 for invoice mistakes made during the past five years. However, it's not guaranteed that TMI will receive the credit in 2015. Jon also thinks TMI might be able to receive a discount postage rate for high-volume campaigns. Finally, Robert tells you he has been in conversations with Delaware Inc. to do additional mailings in October, November, and December 2015.

You will use Excel to determine how different scenarios would affect TMI's 2015 marketing forecast. Your scenarios will focus on the following amounts and the answer to an important question:

- 2015 total costs
- 2015 mail volume
- 2015 average cost per mailing
- 2015 year-end stock amount of Jumbo envelopes
- 2015 year-end stock amount of 6 × 9 envelopes
- 2015 year-end stock amount of #10 envelopes
- Will you have enough envelopes to finish 2015?

In summary, your decision support system (DSS) will include the following inputs:

- Mail volume required to earn a postage discount from the USPS
- Whether the postage credit will be received in 2015
- Additional mail volume from Delaware Inc. if the company approves additional mailings in October, November, and December 2015

Your DSS model must account for the effects of the preceding three inputs on costs, inventory, and mail volumes. If you design the model well, it will let you develop "what-if" scenarios with all the inputs, see the results, and show how they fit into budget requests from the Marketing department.

ASSIGNMENT 1: CREATING A SPREADSHEET FOR DECISION SUPPORT

In this assignment, you will produce a spreadsheet that models TMI's marketing budget for 2015. In Assignment 2, you will write a memorandum that documents your analysis and findings. In Assignment 3, you will prepare and give an oral presentation of your analysis and conclusions to the CEO and chief marketing officer.

First, you will create the spreadsheet model of the budget. The model's budget covers one year (2015). This section helps you set up each of the following spreadsheet components before entering cell formulas:

- Constants
- Inputs
- Summary of Key Results
- Calculations

A discussion of each section follows. The spreadsheet skeleton for this case is available for you to use; it will save you time. To access the spreadsheet skeleton, go to your data files, select Case 6, and then select **TMI.xlsx**.

Constants Section

Your spreadsheet should include the constants shown in Figures 6-1 and 6-2. An explanation of the line items follows each figure.

CONSTANTS		2015
Jumbo Envelope Cost	$	0.89
6x9 Envelope Cost	$	0.60
#10 Envelope Cost	$	0.36
First Class Postage Presorted	$	0.34
First Class Postage Unsorted	$	0.49
Jumbo Handling Cost		$250
6x9 Handling Cost		$150
#10 Handling Cost		$75
Discount Rate		5.0%
Postage Credit Amount from USPS	$	25,000
Envelope Stock as of 1/1/15		
Jumbo		35,000
6x9		97,000
#10		86,000

Source: Microsoft product screenshots used with permission from Microsoft Corporation

FIGURE 6-1   Constants section

- Jumbo Envelope Cost—The unit cost per envelope to the Marketing department.
- 6 × 9 Envelope Cost—The unit cost per envelope to the Marketing department.
- #10 Envelope Cost—The unit cost per envelope to the Marketing department.
- First Class Postage Presorted—The cost of a single mailing at a presorted rate. "Presorted" means that the mailing addresses on the envelopes have been verified and are free of errors. This rate only applies to mail volumes of 5,000 pieces or more.
- First Class Postage Unsorted—The cost of a single mailing at an unsorted rate. Addresses are not verified and may contain errors. This rate only applies to mail volumes of less than 5,000 pieces.
- Jumbo Handling Cost—The costs associated with using Jumbo envelopes in any mailing. The costs usually include those for printer setup and ink.
- 6 × 9 Handling Cost—The costs associated with using 6 × 9 envelopes in any mailing. The costs usually include those for printer setup and ink.
- #10 Handling Cost—The costs associated with using #10 envelopes in any mailing. The costs usually include those for printer setup and ink.
- Discount Rate—The percentage discount for high-volume mailing campaigns.
- Postage Credit Amount from USPS—The credit amount retroactively requested from the USPS based on past invoice mistakes and reconciliations.
- Envelope Stock as of 1/1/15—Starting inventory levels in 2015 for each type of envelope.

The second part of the Constants section lists volume information for the key direct-mail campaigns TMI has planned for 2015.

Expected Campaign Volumes	Jan-15	Feb-15	Mar-15	Apr-15	May-15	Jun-15	Jul-15	Aug-15	Sep-15	Oct-15
ABC Corp. Jumbo										
ABC Corp. 6x9										
Delaware Inc. 6x9										
Delaware Inc. #10										
Blue Hens Fan Club #10										

Source: Microsoft product screenshots used with permission from Microsoft Corporation

FIGURE 6-2   Constants section, continued

- ABC Corp. Jumbo—The campaign requires 5,000 mail pieces in January, March, May, July, September, and November.
- ABC Corp. 6 × 9—The campaign requires 5,000 mail pieces in January, March, May, July, September, and November.

- Delaware Inc. 6 × 9—The campaign requires 30,000 mail pieces in January and June.
- Delaware Inc. #10—The campaign requires 20,000 mail pieces in January and June.
- Blue Hens Fan Club #10—The campaign requires 2,500 mailings each month for the entire year.

Inputs Section

As Jon explained earlier, TMI must answer three important questions for the upcoming year. First, TMI has been in negotiations with the USPS to receive a volume discount. You know that the discount will be 5 percent of postage costs, but you don't know how many mailings are required per campaign to obtain the discount. Also, to correct past invoice mistakes from the USPS, the company is expecting to receive a $25,000 postage credit, possibly in 2015. Finally, your model must account for the possibility of additional mail volume for the Delaware Inc. campaign.

Your spreadsheet should include the following inputs, as shown in Figure 6-3.

INPUTS		Oct-15	Nov-15	Dec-15
Volume Required For Discount		NA	NA	NA
Postage Credit From USPS ("Y" = Yes, "N" = No)	NA	NA	NA	NA
Additional Delaware Inc Campaign	NA			

Source: Microsoft product screenshots used with permission from Microsoft Corporation

FIGURE 6-3   Inputs section

- Volume Required For Discount—The threshold amount of mailings required by the USPS to obtain the postage discount.
- Postage Credit From USPS ("Y" = Yes, "N" = No)—A value to represent the expected response from the USPS.
- Additional Delaware Inc. Campaign—Starting in October 2015, the mail volume for each additional month requested by Delaware Inc. using 6 × 9 envelopes. Be sure that the Expected Campaign Volumes boxes in the Constants section reflect the changing volume.

Summary of Key Results Section

Your spreadsheet should include the results shown in Figure 6-4. An explanation of each item follows the figure.

SUMMARY OF KEY RESULTS	Year End
Yearly Cost Grand Total	$ -
Yearly Mail Volume	-
Average Cost Per Piece	$ -
Year End Jumbo Envelope Inventory	35,000
Year End 6x9 Envelope Inventory	97,000
Year End #10 Envelope Inventory	86,000
Year End Reorder Jumbo Envelopes?	No
Year End Reorder 6x9 Envelopes?	No
Year End Reorder #10 Envelopes?	No

Source: Microsoft product screenshots used with permission from Microsoft Corporation

FIGURE 6-4   Summary of Key Results section

At the end of the year, your spreadsheet should include values for the yearly cost grand total, yearly mail volume, and average cost per piece of mail; year-end inventory levels for each of the three envelopes; and values for whether the company will need to reorder envelopes. The Yearly Cost Grand Total value should be formatted as currency with no decimals, and the Average Cost Per Piece should be formatted as currency with no decimals. The inventory cells should be formatted numbers with no decimals, and the reorder cells should be formatted as text. All of these values are computed elsewhere in the spreadsheet and should be echoed here.

Calculations Section

To create an accurate forecasting tool, you should calculate intermediate results that will be used to determine the year-end numbers needed for budgeting. The calculations shown in Figures 6-5, 6-6, 6-7, and 6-8 are based on expected monthly values for 2015. When called for, use absolute referencing properly. Values must be computed by cell formula; hard-code numbers in formulas only when you are told to do so. Cell formulas should not reference a cell with a value of "NA," which stands for "not applicable."

An explanation of each item in this section follows the figure in which the item is shown.

Inventory	Jan-15	Feb-15	Mar-15	Apr-15	May-15	Jun-15	Jul-15	Aug-15	Sep-15	Oct-15
Total Jumbo Envelopes Used	-	-	-	-	-	-	-	-	-	-
Total 6x9 Envelopes Used	-	-	-	-	-	-	-	-	-	-
Total #10 Envelopes Used	-	-	-	-	-	-	-	-	-	-
Remaining Inventory	Jan-15	Feb-15	Mar-15	Apr-15	May-15	Jun-15	Jul-15	Aug-15	Sep-15	Oct-15
Remaining Jumbo Envelopes	35,000	35,000	35,000	35,000	35,000	35,000	35,000	35,000	35,000	35,000
Remaining 6x9 Envelopes	97,000	97,000	97,000	97,000	97,000	97,000	97,000	97,000	97,000	97,000
Remaining #10 Envelopes	86,000	86,000	86,000	86,000	86,000	86,000	86,000	86,000	86,000	86,000
Reorder Jumbo Envelopes?	No	No	No	No	No	No	No	No	No	No
Reorder 6x9 Envelopes?	No	No	No	No	No	No	No	No	No	No
Reorder #10 Envelopes?	No	No	No	No	No	No	No	No	No	No

FIGURE 6-5   Inventory section of calculations

- Total Jumbo Envelopes Used—The sum of all Jumbo envelopes used per month for all campaigns. Format cells for numbers with no decimals.
- Total 6 × 9 Envelopes Used—The amount of 6 × 9 envelopes used per month for all campaigns. Format cells for numbers with no decimals.
- Total #10 Envelopes Used—The amount of #10 envelopes used per month for all campaigns. Format cells for numbers with no decimals.
- Remaining Jumbo Envelopes—The amount of Jumbo envelopes remaining at the end of each month. For January, this value would be the amount remaining as of January 1, 2015, minus the amount used in January. For the rest of the year, this value would be the amount remaining at the end of the previous month minus the amount used in the current month. Format cells for numbers with no decimals.
- Remaining 6 × 9 Envelopes—The amount of 6 × 9 envelopes remaining at the end of each month. For January, this value would be the amount remaining as of January 1, 2015, minus the amount used in January. For the rest of the year, this value would be the amount remaining at the end of the previous month minus the amount used in the current month. Format cells for numbers with no decimals.
- Remaining #10 Envelopes—The amount of #10 envelopes remaining at the end of each month. For January, this value would be the amount remaining as of January 1, 2015, minus the amount used in January. For the rest of the year, this value would be the amount remaining at the end of the previous month minus the amount used in the current month. Format cells for numbers with no decimals.
- Reorder Jumbo Envelopes?—When the remaining number of Jumbo envelopes is less than or equal to zero, this cell will display "Yes." Otherwise, the cell will display "No." Format the cell as text.
- Reorder 6 × 9 Envelopes?—When the remaining number of 6 × 9 envelopes is less than or equal to zero, this cell will display "Yes." Otherwise, the cell will display "No." Format the cell as text.
- Reorder #10 Envelopes?—When the remaining number of #10 envelopes is less than or equal to zero, this cell will display "Yes." Otherwise, the cell will display "No." Format the cell as text.

Costs Envelopes	Jan-15	Feb-15	Mar-15	Apr-15	May-15	Jun-15	Jul-15	Aug-15	Sep-15
ABC Corp. Jumbo	$ -	$ -	$ -	$ -	$ -	$ -	$ -	$ -	$ -
ABC Corp. 6x9	$ -	$ -	$ -	$ -	$ -	$ -	$ -	$ -	$ -
Delaware Inc. 6x9	$ -	$ -	$ -	$ -	$ -	$ -	$ -	$ -	$ -
Delaware Inc. #10	$ -	$ -	$ -	$ -	$ -	$ -	$ -	$ -	$ -
Blue Hens Fan Club #10	$ -	$ -	$ -	$ -	$ -	$ -	$ -	$ -	$ -
Campaign Postage	Jan-15	Feb-15	Mar-15	Apr-15	May-15	Jun-15	Jul-15	Aug-15	Sep-15
ABC Corp. Jumbo	$ -	$ -	$ -	$ -	$ -	$ -	$ -	$ -	$ -
ABC Corp. 6x9	$ -	$ -	$ -	$ -	$ -	$ -	$ -	$ -	$ -
Delaware Inc. 6x9	$ -	$ -	$ -	$ -	$ -	$ -	$ -	$ -	$ -
Delaware Inc. #10	$ -	$ -	$ -	$ -	$ -	$ -	$ -	$ -	$ -
Blue Hens Fan Club #10	$ -	$ -	$ -	$ -	$ -	$ -	$ -	$ -	$ -
Handling	Jan-15	Feb-15	Mar-15	Apr-15	May-15	Jun-15	Jul-15	Aug-15	Sep-15
ABC Corp. Jumbo	$ -	$ -	$ -	$ -	$ -	$ -	$ -	$ -	$ -
ABC Corp. 6x9	$ -	$ -	$ -	$ -	$ -	$ -	$ -	$ -	$ -
Delaware Inc. 6x9	$ -	$ -	$ -	$ -	$ -	$ -	$ -	$ -	$ -
Delaware Inc. #10	$ -	$ -	$ -	$ -	$ -	$ -	$ -	$ -	$ -
Blue Hens Fan Club #10	$ -	$ -	$ -	$ -	$ -	$ -	$ -	$ -	$ -
Total Campaign Costs	Jan-15	Feb-15	Mar-15	Apr-15	May-15	Jun-15	Jul-15	Aug-15	Sep-15
ABC Corp. Jumbo	$ -	$ -	$ -	$ -	$ -	$ -	$ -	$ -	$ -
ABC Corp. 6x9	$ -	$ -	$ -	$ -	$ -	$ -	$ -	$ -	$ -
Delaware Inc. 6x9	$ -	$ -	$ -	$ -	$ -	$ -	$ -	$ -	$ -
Delaware Inc. #10	$ -	$ -	$ -	$ -	$ -	$ -	$ -	$ -	$ -
Blue Hens Fan Club #10	$ -	$ -	$ -	$ -	$ -	$ -	$ -	$ -	$ -

Source: Microsoft product screenshots used with permission from Microsoft Corporation

FIGURE 6-6   Costs section of calculations

- Envelopes
  - ABC Corp. Jumbo—The total cost of Jumbo envelopes for the ABC Corp. campaign. Format the cell for currency with no decimals.
  - ABC Corp. 6×9—The total cost of 6×9 envelopes for the ABC Corp. campaign. Format the cell for currency with no decimals.
  - Delaware Inc. 6×9—The total cost of 6×9 envelopes for the Delaware Inc. campaign. Format the cell for currency with no decimals.
  - Delaware Inc. #10—The total cost of #10 envelopes for the Delaware Inc. campaign. Format the cell for currency with no decimals.
  - Blue Hens Fan Club #10—The total cost of #10 envelopes for the Blue Hens Fan Club campaign. Format the cell for currency with no decimals.
- Campaign Postage
  - ABC Corp. Jumbo—The total cost of postage for mail sent in Jumbo envelopes for the ABC Corp. campaign. The model should account for whether the mailing qualifies for presorted postage rates. Format the cell for currency with no decimals.
  - ABC Corp. 6×9—The total cost of postage for mail sent in 6×9 envelopes for the ABC Corp. campaign. The model should account for whether the mailing qualifies for presorted postage rates. Format the cell for currency with no decimals.
  - Delaware Inc. 6×9—The total cost of postage for mail sent in 6×9 envelopes for the Delaware Inc. campaign. Format the cell for currency with no decimals.
  - Delaware Inc. #10—The total cost of postage for mail sent in #10 envelopes for the Delaware Inc. campaign. The model should account for instances in which there is no volume. Format the cell for currency with no decimals.
  - Blue Hens Fan Club #10—The total cost of postage for mail sent in #10 envelopes for the Blue Hens Fan Club campaign. The model should account for instances in which there is no volume. Format the cell for currency with no decimals.
- Handling
  - ABC Corp. Jumbo—The handling cost of Jumbo envelopes for the ABC Corp. campaign. The model should account for whether the mailing has no volume. Format the cell for currency with no decimals.

- ABC Corp. 6 × 9—The handling cost of 6 × 9 envelopes for the ABC Corp. campaign. The model should account for whether the mailing qualifies for presorted postage rates. Format the cell for currency with no decimals.
- Delaware Inc. 6 × 9—The handling cost of 6 × 9 envelopes for the Delaware Inc. campaign. Format the cell for currency with no decimals.
- Delaware Inc. #10—The handling cost of #10 envelopes for the Delaware Inc. campaign. The model should account for instances in which there is no volume. Format the cell for currency with no decimals.
- Blue Hens Fan Club #10—The handling cost of #10 envelopes for the Blue Hens Fan Club campaign. The model should account for instances in which there is no volume. Format the cell for currency with no decimals.

- Total Campaign Costs
  - ABC Corp. Jumbo—The sum of all costs of using Jumbo envelopes for the ABC Corp. campaign. Format the cell for currency with no decimals.
  - ABC Corp. 6 × 9—The sum of all costs of using 6 × 9 envelopes for the ABC Corp. campaign. Format the cell for currency with no decimals.
  - Delaware Inc. 6 × 9—The sum of all costs of using 6 × 9 envelopes for the Delaware Inc. campaign. Format the cell for currency with no decimals.
  - Delaware Inc. #10—The sum of all costs of using #10 envelopes for the Delaware Inc. campaign. Format the cell for currency with no decimals.
  - Blue Hens Fan Club #10—The sum of all costs of using #10 envelopes for the Blue Hens Fan Club campaign. Format the cell for currency with no decimals.

	Jan-15	Feb-15	Mar-15	Apr-15	May-15	Jun-15	Jul-15	Aug-15	Sep-15
Monthly Grand Total (Minus Applicable Discount)	$ -	$ -	$ -	$ -	$ -	$ -	$ -	$ -	$ -
Monthly Total Mail Pieces	-	-	-	-	-	-	-	-	-
Monthly Average Cost Per Mail Piece	$ -	$ -	$ -	$ -	$ -	$ -	$ -	$ -	$ -

Source: Microsoft product screenshots used with permission from Microsoft Corporation

FIGURE 6-7 Monthly totals section of calculations

- Monthly Grand Total (Minus Applicable Discount)—The sum of all campaign costs for each month, including postage, envelopes, and handling costs. If a high-volume discount is applicable, it should be calculated here as well. Keep in mind that two conditions must be met before the discount can be calculated: The discount threshold must be greater than zero and the Monthly Total Mail Pieces value must be greater than the threshold.
- Monthly Total Mail Pieces—The sum of all envelopes used in all campaigns for each month.
- Monthly Average Cost Per Mail Piece—The average cost per unit mailed after all costs and volumes have been calculated. In other words, this value is the Monthly Grand Total divided by the Monthly Total Mail Pieces.

Yearly Grand Total	$ -
Yearly Total Mail Volume	-
Yearly Average Cost Per Mail Piece	$ -

Source: Microsoft product screenshots used with permission from Microsoft Corporation

FIGURE 6-8 Yearly totals section of calculations

- Yearly Grand Total—The sum of monthly costs for the year. If the USPS invoice reconciliation efforts were successful in 2015, this cell should include the amount of the credit.
- Yearly Total Mail Volume—The sum of monthly mail volumes for the year.
- Yearly Average Cost Per Mail Piece—The average cost per unit mailed after all costs and volumes have been calculated. In other words, this value is the Yearly Grand Total divided by the Yearly Total Mail Volume.

ASSIGNMENT 2: USING THE SPREADSHEET FOR DECISION SUPPORT

Complete the case by (1) using the spreadsheet to answer Jon's questions and (2) documenting your findings in a memorandum.

The department wants to minimize the funds it needs to request from Finance, so you and TMI's management are interested in the following four test scenarios. The scenario names are based on the possible outcomes.

1. Optimist—Assume that the threshold required for a high-volume discount is 30,000 units, the USPS credits you for old invoices, and Delaware Inc. decides to mail 10,000 units a month during October, November, and December.

2. Optimist Low—Assume that the threshold required for a high-volume discount is 45,000 units, the USPS credits you for old invoices, and Delaware Inc. decides to mail 5,000 units a month during October, November, and December.

3. Normal—Assume that the threshold required for a high-volume discount is 45,000 units, the USPS doesn't credit you for old invoices, and Delaware Inc. will not request additional direct-mail campaigns.

4. Pessimist—Assume that the threshold required for a high-volume discount is 100,000 units, the USPS doesn't credit you for old invoices, and Delaware Inc. will not request additional direct-mail campaigns.

Management will need to know the range of funds it should request from Finance for the upcoming year.

Assignment 2A: Using the Spreadsheet to Gather Data

You have built the spreadsheet to model several possible situations. For each of the four test scenarios, you want to know the yearly costs and volumes and whether you will need to order more envelopes at the end of the year.

You will run "what-if" scenarios with the four sets of input values using Scenario Manager. (See Tutorial C for details on using Scenario Manager.) Set up the four scenarios. Your instructor may ask you to use conditional formatting to make sure your input values are proper. Note that in Scenario Manager you can enter noncontiguous cell ranges, such as C19, D19, C20:F20.

The relevant output cells are Yearly Cost Grand Total, Yearly Mail Volume, Average Cost Per Piece, the year-end inventory levels, and the cells that report whether you need to reorder envelopes at the end of the year. All of these cells are shown in the Summary of Key Results section. Run Scenario Manager to gather the data in a report. When you finish, print the spreadsheet with the input for any of the scenarios, print the Scenario Manager summary sheet, and then save the spreadsheet file a final time.

Assignment 2B: Documenting Your Results in a Memo

Use Microsoft Word to write a brief memo that documents your analysis and results. You can address the memo to the CEO of TMI, Robert Fleckburg. Observe the following requirements:

- Set up your memo as described in Tutorial E.
- In the first paragraph, briefly state the business situation and the purpose of your analysis.
- Next, describe the four scenarios tested and indicate how well TMI fared in each.
- State your conclusions. Do any scenarios indicate a need for management action?
- Support your statements graphically, as your instructor requires. Your instructor may ask you to return to Excel and copy the results of the Scenario Manager summary sheet into the memo. You should include a summary table built in Word based on the Scenario Manager summary sheet results. (This procedure is described in Tutorial E.)

Your table should have the format shown in Figure 6-9.

	Optimist	Optimist Low	Normal	Pessimist
Yearly Cost Grand Total				
Yearly Mail Volume				
Reorder Jumbo?				
Reorder 6 × 9?				
Reorder #10?				

Source: © 2015 Cengage Learning®

FIGURE 6-9   Format of table to insert in memo

ASSIGNMENT 3: GIVING AN ORAL PRESENTATION

Your instructor may ask you to explain your analysis and results in an oral presentation. If so, assume that TMI's management wants the presentation to last 10 minutes or less. Use visual aids or handouts that you think are appropriate. See Tutorial F for tips on preparing and giving an oral presentation.

DELIVERABLES

Your completed case should include the following deliverables for your instructor:

1. A printed copy of your memo
2. Printouts of your spreadsheet and scenario summary
3. Electronic copies of your memo and Excel DSS model

THE VERY BIG BANK STRESS TEST

Decision Support Using Microsoft Excel

PREVIEW

Banks are periodically "stress tested" to assess their financial soundness. In this case, you will use Microsoft Excel to perform a stress test on a bank.

PREPARATION

- Review spreadsheet concepts discussed in class and in your textbook.
- Complete any exercises that your instructor assigns.
- Complete any part of Tutorial C that your instructor assigns. You may need to review the use of If statements and the section called "Cash Flow Calculations: Borrowing and Repayments."
- Review file-saving procedures for Windows programs.
- Refer to Tutorials E and F as necessary.

BACKGROUND

Bank regulators want to know that banks can survive difficult economic conditions and continue to function. Therefore, regulators administer financial "stress tests" as an early warning system. A stress test simulates the way a bank's finances would stand up to challenging conditions and reveals financial weaknesses so they can be remedied. Large banks have an internal testing program in place so they can monitor their own status and be ready when regulators show up unannounced, stress test in hand.

As preparation for a stress test, a review of bank finances is in order. Figure 7-1 shows a bank's possible balance sheet:

Assets		Liabilities and Owner's Equity	
Account	**Amount**	**Account**	**Amount**
Cash	$100,000	Deposits	$300,000
Loans	$400,000	Short-term debt	$100,000
Trading	$300,000	Long-term debt	$300,000
Total	$800,000	Equity	$100,000
		Total	$800,000

Source: © 2015 Cengage Learning®

FIGURE 7-1 Possible bank balance sheet

An explanation of the balance sheet data follows. Amounts shown are for illustrative purposes only.

- Cash—This value is the amount of cash on hand.
- Loans—The bank makes loans to its individual and corporate customers, who promise to pay off the loans in installments. Payments include a reduction on the loan amount owed and interest on the amount owed. Interest received is income to the bank.
- Trading—Nowadays, banks are active players in the securities markets. This means that banks trade in stocks and bonds, which they try to sell at a profit. Banks trade in more exotic securities as well;

sometimes these securities are called derivatives. Trading assets are stocks, bonds, and other securities held for potential profit. Banks have become adept at this kind of investing in recent years. However, the practice remains risky and losses can occur, especially in poor economic times.

- Deposits—People leave their money with the bank "on deposit." The bank pays depositors interest on their money; the interest is an expense to the bank. Deposits are a liability because the bank owes the money to the depositor, who can withdraw the money at any time.
- Short-term debt—Banks borrow money for short terms—sometimes as briefly as overnight. Interest paid is an expense to the bank.
- Long-term debt—Banks borrow money over long terms as well, typically by issuing bonds. Again, interest paid is an expense to the bank.
- Equity—Banks issue preferred and common stock as evidence of ownership. Net income retained in the business (in other words, not paid in dividends) is also part of owner's equity.

Figure 7-2 shows a bank's possible income statement. Amounts shown are for illustrative purposes only.

Account	Amount
Income	
Interest on consumer loans	$10,000
Interest on corporate loans	$20,000
Trading account revenue	$20,000
Total Revenue	$50,000
Expenses	
Interest paid on deposits	$5,000
Interest paid on short-term debt	$6,000
Interest paid on long-term debt	$7,000
Trading account expenses	$5,000
Loan losses	$4,000
Total Expenses	$27,000
Net income before taxes	$23,000
Income tax expense	$3,000
Net income after taxes	$20,000

Source: © 2015 Cengage Learning®

FIGURE 7-2   Possible bank income statement

An explanation of the income statement data follows:

- Interest on consumer and corporate loans—Loans are made to individual customers and to companies. Payments include interest on loan principal still owed.
- Trading account revenue—Stocks, bonds, and other securities are bought and sold. When sold, the proceeds received are income.
- Interest paid on deposits—Depositors are paid for leaving their money with the bank. Interest paid is an expense to the bank.
- Interest paid on short-term and long-term debt—Banks must pay interest to holders of their short-term and long-term debt obligations. The interest paid is an expense.
- Trading account expenses—The cost of the staff and other overhead associated with the trading program are expenses to the bank. The initial cost of the securities traded must be expensed when sold.
- Loan losses—Debtors sometimes cannot make payments. If the loan will never be repaid, it must be written off as a loss; in other words, the loan is no longer an asset of the bank.

You work in the internal testing program of a very large bank in one of our nation's financial centers. You have designed a stress test for the bank that will be administered in Excel. The bank's actual financial data will be used, and it will be supplemented by some assumed values in future years. If the bank remains profitable under the terms of the test, even under assumed adverse conditions, the bank passes the test. If the bank does not remain profitable and therefore fails the test, bank managers will have to consider remedial protective measures.

ASSIGNMENT 1: CREATING A SPREADSHEET FOR DECISION SUPPORT

In this assignment, you will produce a spreadsheet that models the bank's stress test. In Assignment 2, you will write a memorandum that documents your analysis and findings. In Assignment 3, you will prepare and give an oral presentation of your analysis and conclusions to the bank's management.

First, you will create the spreadsheet model of the test. The model's test covers the three years from 2016 to 2018. This section helps you set up each of the following spreadsheet components before entering cell formulas:

- Constants
- Inputs
- Summary of Key Results
- Calculations
- Income Statement and Cash Flow Statement
- Short Term Debt Owed
- Final Stress Test Computations

A discussion of each section follows. The spreadsheet skeleton for this case is available for you to use; it will save you time. To access this skeleton file, go to your data files, select Case 7, and then select **StressTest.xlsx**. In the spreadsheet, all numbers are in millions.

Constants Section

Your spreadsheet should include the constants shown in Figure 7-3. An explanation of the line items follows the figure.

	A	B	C	D	E
1	**THE VERY BIG BANK STRESS TEST**				
2	**($ in millions)**				
3	**CONSTANTS**	**2015**	**2016**	**2017**	**2018**
4	Tax Rate	NA	10%	10%	10%
5	Cash Needed To Start Year	NA	$ 50,000	$ 50,000	$ 50,000
6	Deposits	NA	$ 975,000	$ 990,000	$ 1,000,000
7	Trading Account Assets	NA	$ 200,000	$ 210,000	$ 220,000
8	Long Term Debt Owed	NA	$ 220,000	$ 240,000	$ 260,000
9	Base Equity for Test Purposes	NA	$ 120,000	$ 120,000	$ 120,000

Source: Microsoft product screenshots used with permission from Microsoft Corporation

FIGURE 7-3   Constants section

- Tax Rate—The bank pays federal, state, and foreign income taxes. The combined tax rate is expected to be 10 percent each year.
- Cash Needed To Start Year—The bank needs to have at least $50 billion in cash on hand at the beginning of each year. Assume that the bank could borrow money in the short-term debt market at the end of a year in order to begin the new year with the amount needed. (The amount shown in the spreadsheet is correct; all amounts are in millions, so the $50,000 shown represents $50 billion.)
- Deposits—These fields show the amounts assumed to be owed to depositors each year.
- Trading Account Assets—These fields show the assumed values of stocks, bonds, and other securities owned each year.
- Long Term Debt Owed—These fields show the amounts assumed to be owed to bondholders each year.

- Base Equity for Test Purposes—Part of the stress test computes return on equity: the bank's net income divided by its equity. These fields show the assumed equity values each year for this calculation.

Inputs Section

Banks typically make more money when the general economy is good than when it is poor because more loans are made in good times and borrowers are more capable of repaying in good economies. When prices in securities markets fluctuate a great deal, the markets are said to be volatile. High volatility implies high risk and an increased potential for securities losses.

Your spreadsheet should include the following inputs for the years 2016 to 2018, as shown in Figure 7-4.

	A	B	C	D	E
11	**INPUTS**	**2015**	**2016**	**2017**	**2018**
12	Market Volatility ("H" = High, "N" = Normal)	NA		NA	NA
13	Economy ("G" = Good, "O" = OK, "P" = Poor)	NA			

Source: Microsoft product screenshots used with permission from Microsoft Corporation

FIGURE 7-4   Inputs section

- Market Volatility ("H" = High, "N" = Normal)—Enter a value for the assumed volatility of the securities market during the three-year period. The entry applies to all three years in the test.
- Economy ("G" = Good, "O" = OK, "P" = Poor)—Enter a value for the expected state of the economy during each of the three years. You do not have to enter the same value for each year. For example, successive entries of "P," "O," and "G" would imply improving economic activity during the three-year period.

Summary of Key Results Section

Your spreadsheet should include the results shown in Figure 7-5. An explanation of each item follows the figure.

	A	B	C	D	E
15	**SUMMARY OF KEY RESULTS**	**2015**	**2016**	**2017**	**2018**
16	Net Income After Taxes	NA			
17	End-Of-The-Year Cash On Hand	NA			
18	End-Of-The-Year Short Term Debt Owed	NA			
19	Pass / Fail Stress Test	NA			

Source: Microsoft product screenshots used with permission from Microsoft Corporation

FIGURE 7-5   Summary of Key Results section

For each year, your spreadsheet should show net income after taxes, cash on hand at the end of the year, short-term debt owed at the end of the year, and whether the bank passed or failed the stress test for the year. The net income, cash, and debt cells should be formatted as currency with no decimals. The pass/fail cells should be formatted for text. All values are computed elsewhere in the spreadsheet and should be echoed here.

Calculations Section

You should calculate intermediate results that will be used in the income and cash flow statements that follow. Calculations, as shown in Figure 7-6, may be based on expected year-end 2015 values. When called for, use absolute referencing properly. Values must be computed by cell formula; hard-code numbers in formulas only when you are told to do so. Cell formulas should not reference a cell with a value of "NA," which stands for "not applicable."

An explanation of each item in this section follows the figure.

CALCULATIONS	2015	2016	2017	2018
22 Consumer Loans	$ 500,000			
23 Corporate Loans	$ 400,000			
24 Interest Rate on Consumer Loans	NA			
25 Interest Rate on Corporate Loans	NA			
26 Interest Rate on Trading Assets	NA			
27 Interest Rate Paid on Deposits	NA			
28 Interest Rate Paid on Short Term Debt	NA			
29 Interest Rate Paid on Long Term Debt	NA			
30 Trading Account Expense Ratio	NA			
31 Loan Loss Percentage	NA			

Source: Microsoft product screenshots used with permission from Microsoft Corporation

FIGURE 7-6   Calculations section

- Consumer Loans—The amount of consumer loans made by the bank varies with economic conditions. If the economy during the year is good, the value of consumer loans should increase 7 percent over the prior year's value. If the economy during the year is OK, the value of consumer loans should increase 2 percent over the prior year's value. If the economy during the year is poor, the value of consumer loans should decrease 4 percent from the prior year's value. The expected state of the economy in a year is a value from the Inputs section. Cells should be formatted for currency with no decimals.

- Corporate Loans—The amount of corporate loans made by the bank varies with economic conditions. If the economy during the year is good, the value of corporate loans should increase 5 percent over the prior year's value. If the economy during the year is OK, the value of corporate loans should increase 1 percent over the prior year's value. If the economy during the year is poor, the value of corporate loans should decrease 2 percent from the prior year's value. The expected state of the economy in a year is a value from the Inputs section. Cells should be formatted for currency with no decimals.

- Interest Rate on Consumer Loans—Consumer loans are used for cars, home mortgages, personal matters, school costs, and other purposes. Rates on consumer loans can be high. Consumer loans are often short-term, so the portfolio of loans turns over quickly. Consumer loans at the bank are often variable-rate loans and are adjusted at the bank's discretion as the economy changes. The average rate varies with economic conditions. If the economy is expected to be poor, the average yearly interest rate paid on consumer loans should be 10 percent; otherwise, the average rate is expected to be 14 percent. Format cells for percentages.

- Interest Rate on Corporate Loans—Corporate loans are made to companies for business purposes, such as building a factory. Rates on corporate loans at the bank are lower than those for consumer loans. The corporate loan portfolio turns over every few years, and the average rate varies with economic conditions. If the economy is expected to be poor, the average yearly interest rate paid on corporate loans should be 7 percent; otherwise, the average rate is expected to be 8 percent. Format cells for percentages.

- Interest Rate on Trading Assets—Revenue generation of trading assets can be expressed as a percentage. For example, if $100,000 of trading assets are sold at a rate of 10 percent, the trading asset revenue would be $10,000. If market volatility (from the Inputs section) is expected to be high, the interest rate on trading assets should be 10 percent. If market volatility is expected to be normal, the interest rate on trading assets should be 12 percent. Format cells for percentages.

- Interest Rate Paid on Deposits—The bank expects to pay a higher rate for deposits in good economic times than in poor times. If the economy during a year is expected to be good, the average interest rate paid on deposits should be 1.5 percent. If the economy during a year is expected to be OK, the average rate paid on deposits should be 1.0 percent. If the economy during a year is expected to be poor, the average rate paid on deposits should be 0.5 percent (in other words, one-half of one percent). Format cells for percentages.

- Interest Rate Paid on Short Term Debt—If the economy during a year is expected to be good, the average interest rate paid on short-term debt owed should be 1.5 percent. If the economy during a year is expected to be OK, the average rate paid on short-term debt should be 1.0 percent. If the economy during a year is expected to be poor, the average rate paid on short-term debt should be 0.5 percent. Format cells for percentages.

(handwritten margin notes near top: "G= 4.0 / O= 3.5 / P= 3.0")

- Interest Rate Paid on Long Term Debt—The bank has many bond issues outstanding; bonds are regularly paid off and then replaced with new bond issues. If the economy during a year is expected to be good, the average interest rate paid on long-term debt owed should be 4.0 percent. If the economy during a year is expected to be OK, the average rate paid on long-term debt should be 3.5 percent. If the economy during a year is expected to be poor, the average rate paid on long-term debt should be 3.0 percent. Format cells for percentages.

(handwritten margin notes: "H&P= 110 / P+N=90% / allother=60%")

- Trading Account Expense Ratio—Trading account expenses are expressed as a percentage of trading account revenue. For example, if trading account revenue is $20,000 and the expense ratio is 60 percent, the trading account expenses are $12,000 and the trading account gross profit would be $8,000. The expense ratio can exceed 100 percent if there are trading losses in addition to salaries and overhead. If volatility during a year is expected to be high and the economy during that year is expected to be poor, the trading account expense ratio should be 110 percent. If volatility during a year is expected to be normal and the economy during that year is expected to be poor, the trading account expense ratio should be 90 percent. In all other conditions, the expense ratio is expected to be 60 percent. Format cells for percentages.

(handwritten margin notes: "ADD IF / H&P= 25% / P+N=12% / O= 5% / allother= 3%")

- Loan Loss Percentage—This value is the percentage of loans during a year that must be written off and recognized as an expense in that year. The percentage varies with the economy. Bank regulators are convinced that high volatility in the securities market correlates positively with the level of loan write-offs, and is thus a good predictor of what will happen to a loan portfolio during a year. If volatility during a year is expected to be high and the economy during that year is expected to be poor, the loan loss percentage should be 25 percent. If volatility during a year is expected to be normal and the economy during that year is expected to be poor, the loan loss percentage should be 12 percent. If the economy during a year is expected to be OK, the loan loss percentage should be 5 percent, regardless of volatility. In all other cases, the loan loss percentage is expected to be 3 percent. Format cells for percentages.

Income Statement and Cash Flow Statement Section

The forecast for net income and cash flow starts with the cash on hand at the beginning of the year. This value is followed by the income statement and the calculation of cash on hand at year's end. For readability, format cells in this section as currency with no decimals. Values must be computed by cell formula; hard-code numbers in formulas only if you are told to do so. Cell formulas should not reference a cell with a value of "NA." Your spreadsheets should look like those in Figures 7-7 and 7-8. A discussion of each item in the section follows each figure.

	A	B	C	D	E
33	INCOME STATEMENT AND CASH FLOW STATEMENT	2015	2016	2017	2018
34	Beginning-Of-The-Year Cash On Hand	NA			
35					
36	Revenue	NA			
37	Interest on Consumer Loans	NA			
38	Interest on Corporate Loans	NA			
39	Trading Account Revenue	NA			
40	Total Revenue	NA			
41	Expenses	NA			
42	Interest on Deposits	NA			
43	Interest on Short Term Debt	NA			
44	Interest on Long Term Debt	NA			
45	Trading Account Expense	NA			
46	Loan Losses	NA			
47	Total Expenses	NA			
48	Income before Taxes	NA			
49	Income Tax Expense	NA			
50	Net Income After Taxes	NA			

Source: Microsoft product screenshots used with permission from Microsoft Corporation

FIGURE 7-7   Income Statement and Cash Flow Statement section

- Beginning-Of-The-Year Cash On Hand—This value is the cash on hand at the end of the prior year.
- Interest on Consumer Loans—This amount is a function of consumer loans and the interest rate earned on the loans. These values are shown in the Calculations section.
- Interest on Corporate Loans—This amount is a function of corporate loans and the interest rate earned on the loans. These values are shown in the Calculations section.
- Trading Account Revenue—This amount is a function of trading account assets (from the Constants section) and the interest rate earned on trading assets, which is from the Calculations section.
- Total Revenue—This amount is the sum of the interest on consumer loans, interest on corporate loans, and trading account revenue.
- Interest on Deposits—This amount is a function of deposits (from the Constants section) and the interest rate paid on deposits, which is taken from the Calculations section.
- Interest on Short Term Debt—This amount is the product of short-term debt owed at the beginning of the year and the interest rate on short-term debt (from the Calculations section).
- Interest on Long Term Debt—This amount is the product of long-term debt (from the Constants section) and the interest rate on long-term debt, which is taken from the Calculations section.
- Trading Account Expense—This amount is the product of trading account revenue and the trading account expense ratio (from the Calculations section).
- Loan Losses—This amount is the product of the loan loss percentage and the sum of consumer and corporate loans during the year. These values are taken from the Calculations section.
- Total Expenses—This amount is the sum of the interest on deposits, interest on short-term debt, interest on long-term debt, trading account expense, and loan losses.
- Income before Taxes—This amount is the difference between total revenue and total expenses.
- Income Tax Expense—This amount is zero if income before taxes is zero or negative. Otherwise, income tax expense is the product of the year's tax rate and income before taxes. The tax rate is taken from the Constants section.
- Net Income After Taxes—This amount is the difference between income before taxes and income tax expense.

Line items for the year-end cash calculation are shown in Figure 7-8. In the figure, column B represents 2015, column C is for 2016, and so on. Year 2015 values are NA except for the end-of-year cash on hand, which is $300 billion. (The amount shown in the spreadsheet is correct; all amounts are in millions, so the $300,000 shown represents $300 billion.)

Values must be computed by cell formula; hard-code numbers in formulas only when you are told to do so. Cell formulas should not reference a cell with a value of "NA." An explanation of each item follows the figure.

	A	B	C	D	E
52	Net Cash Position (NCP) Before Borrowing and Repayment of Debt (Beginning of Year Cash + Net Income)	NA			
53	Add: Increase in Borrowing	NA			
54	Less: Repayment of Debt	NA			
55	Equals: End-Of-The-Year Cash On Hand	$ 300,000			

Source: Microsoft product screenshots used with permission from Microsoft Corporation

FIGURE 7-8   End-of-year cash on hand section

- Net Cash Position (NCP) Before Borrowing and Repayment of Debt—The NCP at the end of a year equals the cash at the beginning of the year plus the year's net income after taxes.
- Increase in Borrowing—Assume that the bank can borrow enough cash in the short-term market at the end of the year to reach the minimum cash needed to start the next year. If the NCP is less than this minimum, the bank must borrow enough to start the next year with the minimum. Borrowing increases cash on hand, of course.
- Repayment of Debt—If the NCP is more than the minimum cash needed and some short-term debt is owed at the beginning of the year, you must pay off as much short-term debt as possible, but not take cash below the minimum amount required to start the next year. Repayments reduce cash on hand, of course.
- End-Of-The-Year Cash On Hand—This amount is the NCP plus any short-term borrowing and minus any repayments.

Short Term Debt Owed Section

This section is a calculation of short-term debt owed to the bank at year's end, as shown in Figure 7-9. Year 2015 values are NA except for short-term debt owed at the end of the year, which is $60 billion. (The amount shown in the spreadsheet is correct; all amounts are in millions, so the $60,000 shown represents $60 billion.)

Values must be computed by cell formula; hard-code numbers in formulas only when you are told to do so. Cell formulas should not reference a cell with a value of "NA." An explanation of each item follows the figure.

	A	B	C	D	E
57	**SHORT TERM DEBT OWED**	**2015**	**2016**	**2017**	**2018**
58	Beginning-Of-The-Year Short Term Debt Owed	NA			
59	Add: Increase in Borrowing	NA			
60	Less: Repayment of Debt	NA			
61	Equals: End-Of-The-Year Short Term Debt Owed	$ 60,000			

Source: Microsoft product screenshots used with permission from Microsoft Corporation

FIGURE 7-9   Short Term Debt Owed section

• Beginning-Of-The-Year Short Term Debt Owed—Debt owed at the beginning of a year equals the debt owed at the end of the prior year.

• Increase in Borrowing—This amount has been calculated elsewhere and can be echoed to this section. Borrowing increases the amount of debt owed.

• Repayment of Debt—This amount has been calculated elsewhere and can be echoed to this section. Repayments reduce the amount of debt owed.

• End-Of-The-Year Short Term Debt Owed—In 2016 through 2018, this is the amount owed at the beginning of a year, plus borrowing during the year, and minus repayments during the year.

Final Stress Test Computations Section

This section shows whether the bank passed the stress test or failed it. The Return on Base Equity field shows profitability as it relates to the bank's equity base. Passing or failing is related to the return on base equity and short-term debt increases during the year. As a general rule, a bank that makes very little money or loses money and that must increase short-term debt to continue operating has failed the stress test. A discussion of each item in the section follows Figure 7-10.

	A	B	C	D	E
63	**FINAL STRESS TEST COMPUTATIONS**	**2015**	**2016**	**2017**	**2018**
64	Return on Base Equity	NA			
65	Pass / Fail Stress Test	NA			

Source: Microsoft product screenshots used with permission from Microsoft Corporation

FIGURE 7-10   Final Stress Test Computations section

• Return on Base Equity—This amount is net income after taxes for the year divided by base equity for the year. The second value is taken from the Constants section.

• Pass/Fail Stress Test—If the return on base equity is less than 1 percent and short-term borrowing increased during the year, the bank failed the stress test and the word FAIL should appear in the cell. If the return on base equity is less than 1 percent but short-term debt did not increase, the word HOLD should appear in the cell. In all other cases, the bank passed the stress test and PASS should appear in the cell.

ASSIGNMENT 2: USING THE SPREADSHEET FOR DECISION SUPPORT

Complete the case by (1) using the spreadsheet to gather data about possible stress test scenarios and (2) documenting your findings in a memo.

You assume that the bank will pass stress tests associated with good economic conditions, so you and the bank's management are interested in the following four test scenarios. Scenario names are based on volatility and economy input values.

1. Poor-High—Will the bank pass the stress tests if the economy is poor during the year and security market volatility is high?
2. Poor-Normal—Will the bank pass the stress tests if the economy is poor during the year and security market volatility is normal?
3. OK-High—Will the bank pass the stress tests if the economy is OK during the year and security market volatility is high?
4. OK-Normal—Will the bank pass the stress tests if the economy is OK during the year and security market volatility is normal?

Management will need to address stress test failures in one or more scenarios.

Assignment 2A: Using the Spreadsheet to Gather Data

You have built the spreadsheet to model the stress test situation. For each of the four scenarios, you want to know how much short-term debt is owed and the pass-fail judgment for each year from 2016 to 2018.

You will run "what-if" scenarios with the four sets of input values using Scenario Manager. (See Tutorial C for details on using Scenario Manager.) Set up the four scenarios. Your instructor may ask you to use conditional formatting to make sure your input values are proper. Note that in Scenario Manager you can enter noncontiguous cell ranges, such as C19, D19, C20:F20.

The relevant output cells are the end-of-year short-term debt owed and the pass-fail results for 2016 to 2018. These cells are shown in the Summary of Key Results section. Run Scenario Manager to gather the data in a report. When you finish, print the spreadsheet with the input for any of the scenarios, print the Scenario Manager summary sheet, and then save the spreadsheet file a final time.

Assignment 2B: Documenting Your Results in a Memo

Use Microsoft Word to write a brief memo that documents your analysis and results. You can address the memo to bank management. Observe the following requirements:

- Set up your memo as described in Tutorial E.
- In the first paragraph, briefly state the business situation and the purpose of your analysis.
- Next, describe the four scenarios tested and indicate how well the bank did in each.
- State your conclusions. Do any scenarios indicate a need for management action?
- Support your statements graphically, as your instructor requires. Your instructor may ask you to return to Excel and copy the results of the Scenario Manager summary sheet into the memo. You should include a summary table built in Word based on the Scenario Manager summary sheet results. (This procedure is described in Tutorial E.)

Your table should have the format shown in Figure 7-11.

	Poor-High	Poor-Normal	OK-High	OK-Normal
Short-term debt owed				
Pass/hold/fail				

Source: © 2015 Cengage Learning®

FIGURE 7-11   Format of table to insert in memo

ASSIGNMENT 3: GIVING AN ORAL PRESENTATION

Your instructor may ask you to explain your analysis and results in an oral presentation. If so, assume that bank management wants the presentation to last 10 minutes or less. Use visual aids or handouts that you think are appropriate. See Tutorial F for tips on preparing and giving an oral presentation.

DELIVERABLES

Your completed case should include the following deliverables for your instructor:

1. A printed copy of your memo
2. Printouts of your spreadsheet and scenario summary
3. Electronic copies of your memo and Excel DSS model

PART **3**

TUTORIAL D

BUILDING A DECISION SUPPORT SYSTEM USING MICROSOFT EXCEL SOLVER

In Tutorial C, you learned that Decision Support Systems (DSS) are programs used to help managers solve complex business problems. Cases 6 and 7 were DSS models that used Microsoft Excel Scenario Manager to calculate and display financial outcomes given certain inputs, such as economic outlooks and mortgage interest rates. You used the outputs from Scenario Manager to see how different combinations of inputs affected cash flows and income so that you could make the best decision for expanding your business or selecting a technology to develop and market.

Many business situations require models in which the inputs are not limited to two or three choices, but include large ranges of numbers in more than three variables. For such business problems, managers want to know the best or optimal solution to the model. An optimal solution can either maximize an objective variable, such as income or revenues, or minimize the objective variable, such as operating costs. The formula or equation that represents the target income or operating cost is called an objective function. Optimizing the objective function requires the use of constraints (also called constraint equations), which are rules or conditions you must observe when solving the problem. The field of applied mathematics that addresses problem solving with objective functions and constraint equations is called linear programming. Before the advent of digital computers, linear programming required the knowledge of complex mathematical techniques. Fortunately, Excel has a tool called Solver that can compute the answers to optimization problems.

This tutorial has five sections:

1. **Adding Solver to the Ribbon**—Solver is not installed by default with Excel 2013; you must add it to the application. You may need to use Excel Options to add Solver to the Ribbon.
2. **Using Solver**—This section explains how to use Solver. You will start by determining the best mix of vehicles for shipping exercise equipment to stores throughout the country.
3. **Extending the example**—This section tests your knowledge of Solver as you modify the transportation mix to accommodate changes: additional stores to supply and redesign of the product to reduce shipping volume.
4. **Using Solver on a new problem**—In this section, you will use Solver on a new problem: maximizing the profits for a mix of products.
5. **Troubleshooting Solver**—Because Solver is a complex tool, you will sometimes have problems using it. This section explains how to recognize and overcome such problems.

NOTE

If you need a refresher, Tutorial C offers guidance on basic Excel concepts such as formatting cells and using the IF( ) and AND( ) functions.

ADDING SOLVER TO THE RIBBON

Before you can use Solver, you must determine whether it is installed in Excel. Start Excel; if necessary, open a file to gain access to the Ribbon. Next, click the Data tab on the Ribbon. If you see a group on the right side named Analysis that contains Solver, you do not need to install Solver (see Figure D-1).

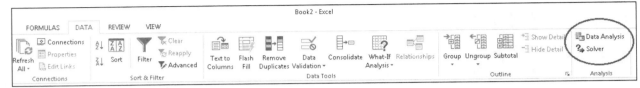

Source: Microsoft product screenshots used with permission from Microsoft Corporation

FIGURE D-1 Analysis group with Solver installed

If the Analysis group or Solver is not shown on the Data tab of the Ribbon, do the following:

1. Click the File tab.
2. Click Options (see Figure D-2).
3. Click Add-Ins (see Figure D-3) to display the available add-ins in the right pane.
4. Click Go at the bottom of the right pane. The window shown in Figure D-4 appears.
5. Click the Solver Add-in box as well as the Analysis ToolPak and Analysis ToolPak-VBA boxes. (You will need the latter options in a subsequent case, so install them now with Solver.)
6. Click OK to close the window and return to the Ribbon. If you click the Data tab again, you should see the Analysis group with Data Analysis and Solver on the right.

Source: Microsoft product screenshots used with permission from Microsoft Corporation

FIGURE D-2 Excel Options selection

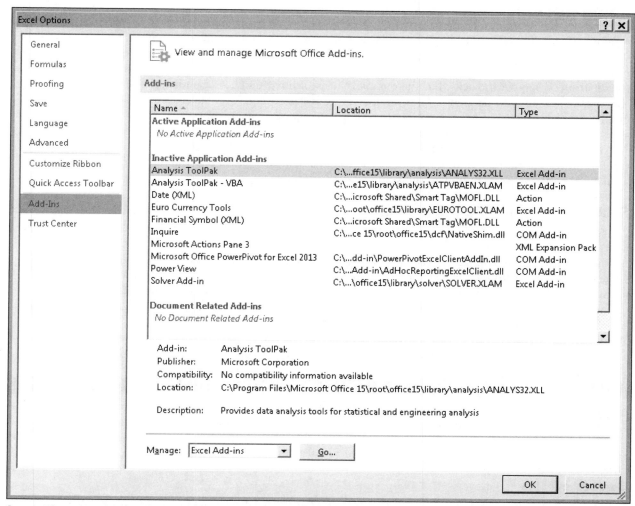

Source: Microsoft product screenshots used with permission from Microsoft Corporation

FIGURE D-3 Add-Ins pane

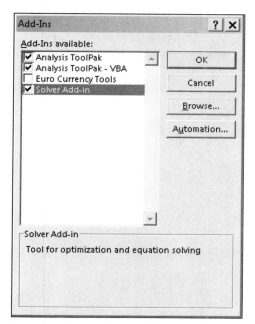

Source: Microsoft product screenshots used with permission from Microsoft Corporation

FIGURE D-4 Add-Ins window with Solver, Analysis ToolPak, and Analysis ToolPak VBA selected

USING SOLVER

A fictional company called CV Fitness builds exercise machines in its plant in Memphis, Tennessee and ships them to its stores across the country. The company has a small fleet of trucks and tractor-trailers to ship its products from the factory to its stores. It costs less money per cubic foot of capacity to ship products with tractor-trailers than with trucks, but the company has a limited number of both types of vehicles and must ship a specified amount of each type of product to each destination. You have been asked to determine the optimal mix of trucks and tractor-trailers to send merchandise to each store. The optimal mix will have the lowest total shipping cost while ensuring that the required quantity of products is shipped to each store.

To use Solver, you must set up a model of the problem, including the factors that can vary (the mix of trucks and tractor-trailers) and the constraints on how much they can vary (the number of each vehicle available). Your goal is to minimize the shipping cost.

Setting Up a Spreadsheet Skeleton

CV Fitness makes three fitness machines: exercise bikes (EB), elliptical cross-trainers (CT), and treadmills (TM). When packaged for shipment, their shipping volumes are 12, 15, and 22 cubic feet, respectively. The finished machines are shipped via ground transportation to five stores in Philadelphia, Atlanta, Miami, Chicago, and Los Angeles. Your vehicle fleet consists of 12 trucks and 6 tractor-trailers. Each truck has a capacity of 1,500 cubic feet, and each tractor-trailer has a capacity of 2,350 cubic feet. The spreadsheet includes the road distances from your plant in Memphis to each store, along with each store's demand for the three fitness machines.

What is the best mix of trucks and tractor-trailers to send to each destination? You will learn how to use Solver to determine the answer. The spreadsheet components are discussed in the following sections.

AT THE KEYBOARD

Start by saving your blank spreadsheet. Use a descriptive filename so you can find it easily later—**CV Fitness Trucking Problem.xlsx** should work well. Then enter the skeleton and formulas as directed in the following sections. *You can also download the spreadsheet skeleton if you prefer; it will save you time.* To access this skeleton file, select Tutorial D from your data files, and then select **CV Fitness Trucking Problem.xlsx**.

Spreadsheet Title

Resize Column A, as illustrated in Figure D-5, to give your spreadsheet a small border on the left side. Enter the spreadsheet title in cell B1. Merge and center cells B1 through F1 using the Merge & Center button in the Alignment group of the Home tab.

Constants Section

Your spreadsheet should have a section for values that will not change. Figure D-5 shows a skeleton of the Constants section and the values you should enter. A discussion of the line items follows the figure.

	A	B	C	D	E	F
1		CV Fitness, Inc. Truck Load Management Problem				
2						
3		Constants Section:				
4			Volume Cu. Ft.	Operating Cost per mi.	Operating Cost per mi-cu. Ft.	Available Fleet
5		Truck	1500	$1.00	$0.000667	12
6		Tractor Trailer	2350	$1.30	$0.000553	6
7						
8		Exercise Bike (EB)	12			
9		Elliptical Crosstrainer (CT)	15			
10		Treadmill (TM)	22			

Source: Microsoft product screenshots used with permission from Microsoft Corporation

FIGURE D-5   Spreadsheet title and Constants section

- In column C, enter the Volume Cu. Ft., which is the cubic-foot capacity of the vehicles as well as the shipping volume for each item of exercise equipment.
- In column D, enter the Operating Cost per mi., which is the cost per mile driven for each type of vehicle.
- In column E, enter the Operating Cost per mi.-cu.ft. This value is actually a formula: the operating cost per mile divided by the vehicle volume in cubic feet. Normally you do not put formulas in the Constants section, but in this case it lets you see the relative cost efficiencies of each vehicle. Assuming that both types of vehicles can be filled to capacity, the tractor-trailer is the preferred vehicle for shipping cost efficiency.
- In column F, enter the values for the Available Fleet, which is the number of each type of vehicle your company owns or leases.

You can update the Constants section as the company adds more products to its offerings or adds vehicles to its fleet.

NOTE

The column headings in the Constants section contain two or three lines to keep the columns from becoming too wide. To create a line break in a cell, hold down the Alt key and press Enter.

Now is a good time to save your workbook again. Keep the name you assigned earlier.

Calculations and Results Section

The structure and format of your Calculations and Results section will vary greatly depending on the nature of the problem you need to solve. In some Solver models, you might need to maximize income, which means you might also have an Income Statement section. In other Solver models, you may want to have a separate Changing Cells section that contains cells Solver will manipulate to obtain a solution. In this tutorial, you want to minimize shipping costs while meeting the product demand of your stores. You can accomplish this task by building a single unified table that includes the distances to the stores, the product demand for each store, and the shipping alternatives and costs.

A unified Calculations and Results section makes sense in this model for several reasons. First, it simplifies writing and copying the formulas for the needed shipping volumes, the vehicle capacity totals, and the shipping costs to each destination. Second, a well-organized table allows you to easily identify the changing cells, which Solver will manipulate to optimize the solution, as well as the total cost (or optimization cell). Finally, a unified table allows your management team to visualize both the problem and its solution.

When creating a complex table, it is often a good idea to sketch the table's structure first to see how you want to organize the data. Format the table structure, then enter the data you are given for the problem. Write the cells that contain the formulas last, starting with all the formulas in the first row. If you do a good job structuring your table, you will be able to copy the first-row formulas to the other rows.

Build the blank table shown in Figure D-6. A discussion of the rows and columns follows the figure.

NOTE

Leave rows 11 and 12 blank between the Constants section and the Calculations and Results section. You then will have room to add an extra product to your Constants section later.

	A	B	C	D	E	F	G	H	I	J	K	L	M	N
12														
13		**Calculations and Results Section:**												
14		**Distance/Demand Table**		**Store Demand**			**Vehicle Loading**							**Cost**
15		**Distance Table (from Memphis Plant)**	**Miles**	**EB**	**CT**	**TM**	**Volume Required**	**Trucks**	**Volume for Trucks**	**Tractor-Trailers**	**Volume for Tractor-Trailers**	**Total Vehicle Capacity**	**% of Vehicle Capacity Utilized**	**Shipping Cost**
16		Philadelphia Store	1010	140	96	86								
17		Atlanta Store	380	76	81	63								
18		Miami Store	1000	56	64	52								
19		Chicago Store	540	115	130	150								
20		Los Angeles Store	1810	150	135	180								
21						Totals:								
22		Fill Legend:			Changing Cells									Total Cost
23					Optimization Cell									

Source: Microsoft product screenshots used with permission from Microsoft Corporation

FIGURE D-6 Blank table for Calculations and Results section

- In row 13, enter "Calculations and Results Section:" as the title of the table.
- In row 14, columns B and C, enter "Distance/Demand Table" as a column heading. Merge and center the heading in the two columns.
- In row 14, columns D, E, and F, enter "Store Demand" as a column heading. Merge and center the heading in the three columns.
- In row 14, columns G through M, enter "Vehicle Loading" as a column heading. Merge and center the heading across the columns.
- In row 14, column N, enter "Cost" as a centered column heading.
- In row 15, column B, enter "Distance Table (from Memphis Plant)" as a centered column heading.
- In row 15, column C, enter "Miles" as a centered column heading.
- In row 15, columns D, E, and F, enter "EB," "CT," and "TM," respectively, as equipment headings.
- In row 15, columns G through N, enter "Volume Required," "Trucks," "Volume for Trucks," "Tractor-Trailers," "Volume for Tractor-Trailers," "Total Vehicle Capacity," "% of Vehicle Capacity Utilized," and "Shipping Cost," respectively, as column headings.
- In rows 16 through 20, column B, enter the destination store locations.
- In rows 16 through 20, column C, enter the number of miles to the destination store locations.
- In rows 16 through 20, columns D through F, enter the number of exercise bikes (EB), cross-trainers (CT), and treadmills (TM) to be shipped to each store location.
- Rows 16 through 20, columns G through N, will contain formulas or "seed values" later. Leave them blank for now, but fill cells H16 through H20 and cells J16 through J20 with a light color to indicate that they are the changing cells for Solver. To fill a cell, use the Fill Color button in the Font group.
- In cell F21, enter "Totals:" to label the following cells in the row.
- Cells G21 through N21 will be used for column totals. Fill cell N21 with a slightly darker shade than you used for the changing cells. Cell N21 is your optimization cell.
- In cell B22, enter "Fill Legend:" as a label.
- Fill cell C22 with the fill color you selected for the changing cells.
- In cells D22 and E22, enter "Changing Cells" as the label for the fill color. Merge and center the label in the cells.
- In cell N22, enter "Total Cost" as the label for the value in cell N21.
- Fill cell C23 with the fill color you selected for the optimization cell.
- In cells D23 and E23, enter "Optimization Cell" as the label for the fill color. Merge and center the label in the cells.

Figure D-7 illustrates a magnified section of the Distance/Demand table in case the numbers in Figure D-6 are difficult to read.

	Distance Table (from Memphis Plant)	Miles	EB	CT	TM
12					
13	**Calculations and Results Section:**				
14	**Distance/Demand Table**		**Store Demand**		
15	**Distance Table (from Memphis Plant)**	**Miles**	**EB**	**CT**	**TM**
16	Philadelphia Store	1010	140	96	86
17	Atlanta Store	380	76	81	63
18	Miami Store	1000	56	64	52
19	Chicago Store	540	115	130	150
20	Los Angeles Store	1810	150	135	180
21					Totals:
22	Fill Legend:		Changing Cells		
23			Optimization Cell		

Source: Microsoft product screenshots used with permission from Microsoft Corporation

FIGURE D-7   Magnified view of the Distance/Demand table

Use the Borders menu in the Font group to select and place appropriate borders around parts of the Calculations and Results section (see Figure D-8). The All Borders and Outside Borders selections are the most useful borders for your table.

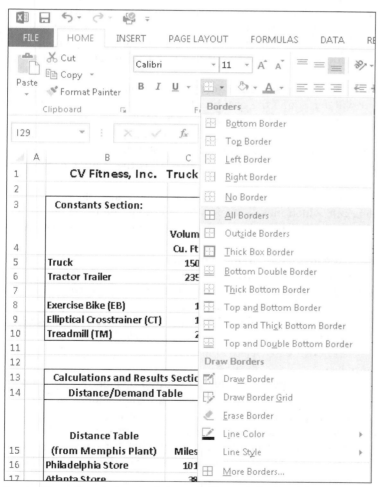

Source: Microsoft product screenshots used with permission from Microsoft Corporation

FIGURE D-8   Borders menu

Next, you write the formulas for the volume and cost calculations. Figure D-9 shows a magnified view of the Vehicle Loading and Cost sections. A discussion of the formulas required for the cells follows the figure.

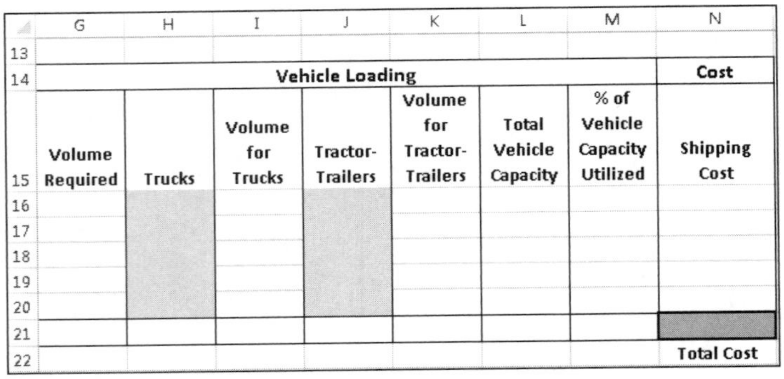

	G	H	I	J	K	L	M	N
13								
14			Vehicle Loading					Cost
15	Volume Required	Trucks	Volume for Trucks	Tractor-Trailers	Volume for Tractor-Trailers	Total Vehicle Capacity	% of Vehicle Capacity Utilized	Shipping Cost
16								
17								
18								
19								
20								
21								
22								Total Cost

Source: Microsoft product screenshots used with permission from Microsoft Corporation

FIGURE D-9   Vehicle Loading and Cost sections

For illustration purposes, the cell numbers in the following list refer to values for the Philadelphia store.

- Volume Required—Cell G16 contains the total shipping volume of the three types of equipment shipped to the Philadelphia store. The formula for this cell is =D16*$C$8+E16*$C$9+F16*$C$10. Cells D16, E16, and F16 are the quantities of each item to be shipped, and cells $C$8, $C$9, and $C$10 are the shipping volumes for the exercise bike, cross-trainer, and treadmill, respectively. When taking values from the Constants section to calculate formulas, you almost always should use absolute cell references ($) because you will copy the formulas down the columns.
- Trucks—Cell H16 contains the number of trucks selected to ship the merchandise. Cell H16 is a changing cell, which means Solver will determine the best number of trucks to use and place the number in this cell. For now, you should "seed" the cell with a value of 1.
- Volume for Trucks—Cell I16 contains the number of trucks selected, multiplied by the capacity of a truck. The capacity value is taken from the Constants section. The formula for this cell is =H16*$C$5. Cell H16 is the number of trucks selected, and cell $C$5 is the volume capacity of the truck in cubic feet.
- Tractor-Trailers—Cell J16 contains the number of tractor-trailers selected to ship the merchandise. Cell J16 is a changing cell, which means Solver will determine the best number of tractor-trailers to use and place the number in this cell. For now, you should "seed" the cell with a value of 1.
- Volume for Tractor-Trailers—Cell K16 contains the number of tractor-trailers selected, multiplied by the capacity of a tractor-trailer. The capacity value is taken from the Constants section. The formula for this cell is =J16*$C$6. Cell J16 is the number of tractor-trailers selected, and cell $C$6 is the cubic feet capacity of the tractor-trailer.
- Total Vehicle Capacity—Cell L16 contains the sum of the Volume for Trucks and the Volume for Tractor-Trailers. The formula for this cell is =I16+K16. You need to know the Total Vehicle Capacity to make sure that you have enough capacity to ship the Volume Required. This value will be one of your constraints in Solver.
- % of Vehicle Capacity Utilized—Cell M16 contains the Volume Required divided by the Total Vehicle Capacity. The formula for this cell is =G16/L16; after entering the formula, format it as a percentage using the % button in the Number group. Although this information is not required to minimize shipping costs, it is useful for managers to know how much space was filled in the selected vehicles. Alternatively, you could run Solver to determine the highest space utilization on the vehicles rather than the lowest cost. Note that you cannot use more than 100% of the available space on the vehicles.

- Shipping Cost—Cell N16 contains the following calculation:

  Mileage to destination store × Number of trucks selected × Cost per mile for trucks + Mileage to destination store × Number of tractor-trailers selected × Cost per mile for tractor-trailers

  The formula for this cell is =H16*C16*$D$5+J16*C16*$D$6. Note that absolute cell references for the cost-per-mile values are taken from the Constants section.

If you entered the formulas correctly in row 16, your table should look like Figure D-10.

	G	H	I	J	K	L	M	N
14	Vehicle Loading							Cost
15	Volume Required	Trucks	Volume for Trucks	Tractor-Trailers	Volume for Tractor-Trailers	Total Vehicle Capacity	% of Vehicle Capacity Utilized	Shipping Cost
16	5012	1	1500	1	2350	3850	130%	$2,323.00
17								
18								
19								
20								
21								
22								Total Cost

Source: Microsoft product screenshots used with permission from Microsoft Corporation

FIGURE D-10   Vehicle Loading and Cost sections with formulas entered in the first row

To complete the empty cells in rows 17 through 20, you can copy the formulas from cells G16 through N16 to the rest of the rows. Click and drag to select cells G16 through N16, then right-click and select Copy from the menu (see Figure D-11).

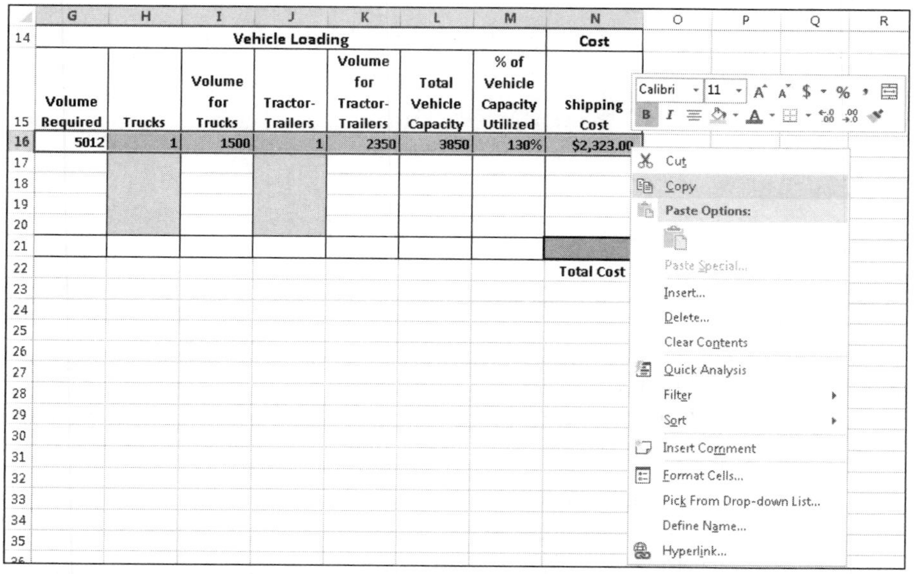

Source: Microsoft product screenshots used with permission from Microsoft Corporation

FIGURE D-11   Copying formulas

Next, select cells G17 through N20, which are in the four rows beneath row 16. Either press Enter or click Paste in the Clipboard group. The formulas from row 16 should be copied to the rest of the destination cities (see Figure D-12).

	G	H	I	J	K	L	M	N
14	Vehicle Loading							Cost
15	Volume Required	Trucks	Volume for Trucks	Tractor-Trailers	Volume for Tractor-Trailers	Total Vehicle Capacity	% of Vehicle Capacity Utilized	Shipping Cost
16	5012	1	1500	1	2350	3850	130%	$2,323.00
17	3513	1	1500	1	2350	3850	91%	$874.00
18	2776	1	1500	1	2350	3850	72%	$2,300.00
19	6630	1	1500	1	2350	3850	172%	$1,242.00
20	7785	1	1500	1	2350	3850	202%	$4,163.00
21								
22								Total Cost

Source: Microsoft product screenshots used with permission from Microsoft Corporation

FIGURE D-12 Formulas from row 16 successfully copied to rows 17 through 20

You have one row of formulas to complete: the Totals row. You will use the AutoSum function to sum up one column, and then copy the formula to the rest of the columns *except* cell M21. This cell is not actually a total, but an overall capacity utilization rate.

To enter the sum of cells G16 through G20 in cell G21, select cells G16 through G21, then click AutoSum in the Editing group on the Home tab of the Ribbon (see Figure D-13).

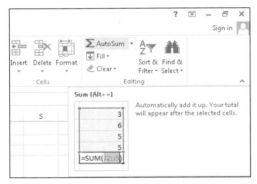

Source: Microsoft product screenshots used with permission from Microsoft Corporation

FIGURE D-13 AutoSum button in the Editing group

Cell G21 should now contain the formula =SUM(G16:G20), and the displayed answer should be 25716. Now you can copy cell G21 to cells H21, I21, J21, K21, L21, and N21. When you have completed this section of the table, it should have the values shown in Figure D-14.

	G	H	I	J	K	L	M	N
14	Vehicle Loading							Cost
15	Volume Required	Trucks	Volume for Trucks	Tractor-Trailers	Volume for Tractor-Trailers	Total Vehicle Capacity	% of Vehicle Capacity Utilized	Shipping Cost
16	5012	1	1500	1	2350	3850	130%	$2,323.00
17	3513	1	1500	1	2350	3850	91%	$874.00
18	2776	1	1500	1	2350	3850	72%	$2,300.00
19	6630	1	1500	1	2350	3850	172%	$1,242.00
20	7785	1	1500	1	2350	3850	202%	$4,163.00
21	25716	5	7500	5	11750	19250		$10,902.00
22								Total Cost

Source: Microsoft product screenshots used with permission from Microsoft Corporation

FIGURE D-14 Totals cells completed

The last formula to enter is for cell M21. This is not a total, but an overall percentage of Vehicle Capacity Utilized for all the vehicles used. This calculation uses the same formula as the cell above it, so you can simply copy cell M20 to cell M21. The formula for this cell is =G21/L21, which is Volume Required divided by Total Vehicle Capacity, expressed as a percentage. Your completed spreadsheet should look like Figure D-15.

	A	B	C	D	E	F	G	H	I	J	K	L	M	N
14		Distance/Demand Table			Store Demand			Vehicle Loading						Cost
15		Distance Table (from Memphis Plant)	Miles	EB	CT	TM	Volume Required	Trucks	Volume for Trucks	Tractor-Trailers	Volume for Tractor-Trailers	Total Vehicle Capacity	% of Vehicle Capacity Utilized	Shipping Cost
16		Philadelphia Store	1010	140	96	86	5012	1	1500	1	2350	3850	130%	$2,323.00
17		Atlanta Store	380	76	81	63	3513	1	1500	1	2350	3850	91%	$874.00
18		Miami Store	1000	56	64	52	2776	1	1500	1	2350	3850	72%	$2,300.00
19		Chicago Store	540	115	130	150	6630	1	1500	1	2350	3850	172%	$1,242.00
20		Los Angeles Store	1810	150	135	180	7785	1	1500	1	2350	3850	202%	$4,163.00
21						Totals:	25716	5	7500	5	11750	19250	134%	$10,902.00
22		Fill Legend:			Changing Cells									Total Cost
23					Optimization Cell									

Source: Microsoft product screenshots used with permission from Microsoft Corporation

FIGURE D-15 Completed Calculations and Results section

Working the Model Manually

Now that you have a working model, you could manipulate the number of trucks and tractor-trailers manually to obtain a solution to the shipping problem. You would need to observe the following rules (or constraints):

1. Assign enough Total Vehicle Capacity to meet the Volume Required for each destination. (In other words, you cannot exceed 100% of Vehicle Capacity Utilized.)
2. The total number of trucks and tractor-trailers you assign cannot exceed the number available in your fleet.

Try to assign your trucks and tractor-trailers to meet your shipping requirements, and note the total shipping costs—you may get lucky and come up with an optimal solution. The tractor-trailers are more cost efficient than the trucks, but the problem is complicated by the fact that you want to achieve the best capacity utilization as well. In some instances, the trucks may be a better fit. Figure D-16 shows a sample solution determined from working the problem manually.

Source: Microsoft product screenshots used with permission from Microsoft Corporation

FIGURE D-16 Manual attempt to solve the vehicle loading problem optimally

This probably looks like a good solution—after all, you have not violated any of your constraints, and you have a 94% average vehicle capacity utilization. But is it the most cost-effective solution for your company? This is where Solver comes in.

Setting Up Solver Using the Solver Parameters Window

To access the Solver pane, click the Data tab on the Ribbon, then click Solver in the Analysis group on the far right side of the Ribbon. The Solver Parameters window appears (see Figure D-17).

NOTE

Solver in Excel 2010 and 2013 has changed significantly from earlier versions of Excel. It allows three different calculation methods, and it allows you to specify an amount of time and number of iterations to perform before Excel ends the calculation. Refer to Microsoft Help for more information.

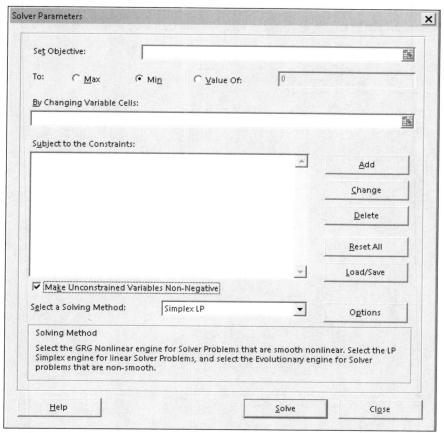

Source: Microsoft product screenshots used with permission from Microsoft Corporation

FIGURE D-17   Solver Parameters window

The Solver Parameters window in Excel 2013 looks intimidating at first. However, to solve linear optimization problems, you have to satisfy only three sets of conditions by filling in the following fields:

- Set Objective—Specify the optimization cell.
- By Changing Variable Cells—Specify the changing cells in your worksheet.
- Subject to the Constraints—Define all of the conditions and limitations that must be met when seeking the optimal solution.

The following sections explain these fields in detail. You may also need to click the Options button and select one or more options for solving the problem. Most of the cases in this book are linear problems, so you can set the solving method to Simplex LP, as shown in Figure D-17. If this method does not work in later cases, you can select the GRG Nonlinear or Evolutionary method to try to solve the problem. Note that the GRG Nonlinear and Evolutionary solving methods are available only in Excel 2010 and 2013.

Optimization Cell and Changing Cells

To use Solver successfully, you must first specify the cell you want to optimize—in this case, the total shipping cost, or cell N21. To fill the Set Objective field, click the button at the right edge of the field, and then click cell N21 in the spreadsheet. You could also type the cell address in the window, but selecting the cell in the spreadsheet reduces your chance of entering the wrong cell address. Next, specify whether you want Solver to seek the maximum or minimum value for cell N21. Because you want to minimize the total shipping cost, click the radio button next to Min.

Next, tell Solver which cell values it will change to determine the optimal solution. Use the By Changing Variable Cells field to specify the range of cells that you want Solver to manipulate. Again, click the button at the right edge of the field, select the cells that contain the numbers of trucks (H16 to H20), and then hold down the

Ctrl key and select the cells that contain the numbers of tractor-trailers (J16 to J20). If you used a fill color for the changing cells, they will be easy to find and select. The Solver Parameters window should look like Figure D-18.

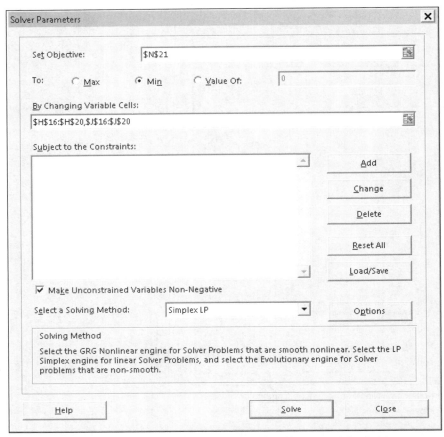

Source: Microsoft product screenshots used with permission from Microsoft Corporation

FIGURE D-18   Solver Parameters window with the objective cell and changing cells entered

Note that Solver has added absolute cell references (the $ signs before the column and row designators) for the cells you have specified. Solver will also add these references to the constraints you define. Solver adds the references to preserve the links to the cells in case you revise the worksheet in the future. In fact, you will make changes to the worksheet later in the tutorial.

Defining and Entering Constraints

For Solver to successfully determine the optimum solution for the shipping problem, you need to specify what constraints or rules it must observe to calculate the solution. Without constraints, Solver theoretically might calculate that the best solution is not to ship anything, resulting in a cost of zero. Furthermore, if you failed to define variables as positive numbers, Solver would select "negative trucks" to maximize "negative costs." Finally, the vehicles are indivisible units—you cannot assign a fraction of a vehicle for a fraction of the cost, so you must define your changing cells as integers to satisfy this constraint.

Aside from the preceding logical constraints, you have operational constraints as well. You cannot assign more vehicles than you have in your fleet, and the vehicles you assign must have at least as much total capacity as your shipping volume.

Before entering the constraints in the Solver Parameters window, it is a good idea to write them down in regular language. You must enter the following constraints for this model:

- All trucks and tractor-trailers in the changing cells must be integers greater than or equal to zero.
- The sums of trucks and tractor-trailers assigned (cells H21 and J21) must be less than or equal to the available trucks and tractor-trailers (cells F5 and F6, respectively).

- The Total Vehicle Capacity for the vehicles assigned to each store (cells L16 to L20) must be greater than or equal to the Volume Required to be shipped to each store (cells G16 to G20, respectively).

You are ready to enter the constraints as equations or inequalities in the Add Constraint window. To begin, click the Add button in the Solver Parameters window. In the window that appears (see Figure D-19), click the button at the right edge of the Cell Reference box, select cells H16 to H20, and then click the button again. Next, click the drop-down menu in the middle field and select > =. Then go to the Constraint field and type 0. Finally, click Add; otherwise, the constraint you defined will not be added to the list defined in the Solver Parameters window.

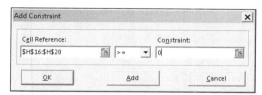

Source: Microsoft product screenshots used with permission from Microsoft Corporation

FIGURE D-19   Add Constraint window

You can continue to add constraints in the Add Constraint window. For this example, enter the constraints shown in the completed Solver Parameters window in Figure D-20. When you finish, click Add to save the last constraint, then click Cancel in the Add Constraint window to return to the Solver Parameters window.

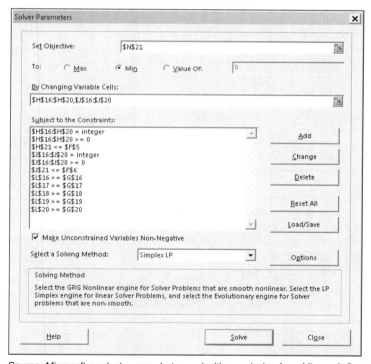

Source: Microsoft product screenshots used with permission from Microsoft Corporation

FIGURE D-20   Completed Solver Parameters window

If you have difficulty reading the constraints listed in Figure D-20, use the following list instead:

- $H$16:$H$20 = integer
- $H$16:$H$20 >= 0
- $H$21 <= $F$5
- $J$16:$J$20 = integer
- $J$16:$J$20 >= 0

- $J$21 <= $F$6
- $L$16 >= $G$16
- $L$17 >= $G$17
- $L$18 >= $G$18
- $L$19 >= $G$19
- $L$20 >= $G$20

You should also click the Options button in the Solver Parameters window and check the Options window shown in Figure D-21. You can use this window to set the maximum amount of time and iterations you want Solver to run before stopping. Make sure that both options are set at 100 for now, but remember that Solver may need more time and iterations for more complex problems. To get the best solution, you should set the Integer Optimality (%) to zero. Click OK to close the window.

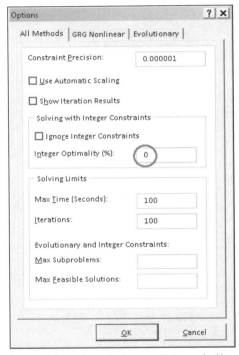

Source: Microsoft product screenshots used with permission from Microsoft Corporation

FIGURE D-21   Solver Options window with Integer Optimality set to zero

You are ready to run Solver to find the optimal solution. Click Solve at the bottom of the Solver Parameters window. Solver might require only a few seconds or more than a minute to run all the possible iterations—the status bar at the bottom of the Excel window displays iterations and possible solutions continuously until Solver finds an optimal solution or runs out of time (see Figure D-22).

Source: Microsoft product screenshots used with permission from Microsoft Corporation

FIGURE D-22   Excel status bar showing Solver running through possible solutions

A new window will appear eventually, indicating that Solver has found an optimal solution to the problem (see Figure D-23). The portion of the spreadsheet that displays the assigned vehicles and shipping cost should be visible below the Solver Results window. Solver has assigned 9 of the 12 trucks and all 6 tractor-trailers, for a total shipping cost of $17,398. The earlier manual attempt to solve the problem (see Figure D-16) assigned all 12 trucks and 4 tractor-trailers, for a total shipping cost of $18,122. Using Solver in this situation saved your company $724.

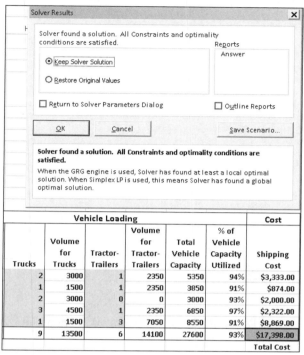

Source: Microsoft product screenshots used with permission from Microsoft Corporation

FIGURE D-23   Solver Results window

If the Solver Results window does not report an optimal solution to the problem, it will report that the problem could not be solved given the changing cells and constraints you specified. For instance, if you had not had enough vehicles in your fleet to carry the required shipping volume to all the destinations, the Solver Results window might have looked like Figure D-24. In the figure, your vehicle fleet was reduced to 10 trucks and 5 tractor-trailers, so Solver could not find a solution that satisfied the shipping volume constraints.

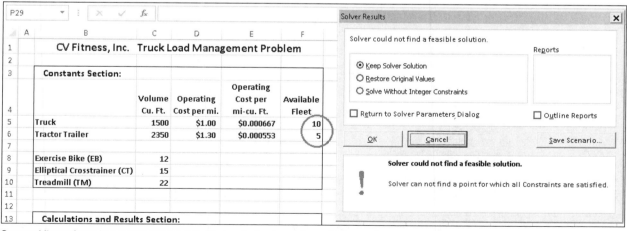

Source: Microsoft product screenshots used with permission from Microsoft Corporation

FIGURE D-24   Solver could not find a feasible solution with a reduced vehicle fleet

Fortunately, Solver did find an optimal solution. To update the spreadsheet with the new optimal values for the changing cells and optimization cell, click OK in the Solver Results window. You can also create an Answer Report by clicking the Answer option in the Solver Results window (see Figure D-25) and then clicking OK.

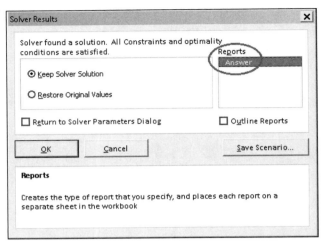

Source: Microsoft product screenshots used with permission from Microsoft Corporation

FIGURE D-25   Creating an Answer Report

Excel will create a report in a separate sheet called Answer Report 1. The Answer Report is shown in Figures D-26 and D-27.

	A	B	C	D	E	F
1	Microsoft Excel 15.0 Answer Report					
2	Worksheet: [CV Fitness Trucking Problem--Working Copy-12th Edition.xlsx]Solver Solution					
3	Report Created: 7/3/2014 12:26:25 PM					
4	Result: Solver found a solution. All Constraints and optimality conditions are satisfied.					
5	Solver Engine					
6	Engine: Simplex LP					
7	Solution Time: 1.092 Seconds.					
8	Iterations: 3 Subproblems: 1416					
9	Solver Options					
10	Max Time 100 sec, Iterations 100, Precision 0.000001					
11	Max Subproblems Unlimited, Max Integer Sols Unlimited, Integer Tolerance 0%, Assume NonNegative					
12						
13	Objective Cell (Min)					
14		Cell	Name	Original Value	Final Value	
15		$N$21	Totals: Shipping Cost	$18,122.00	$17,398.00	
16						
17	Variable Cells					
18		Cell	Name	Original Value	Final Value	Integer
19		$H$16	Philadelphia Store Trucks	2	2	Integer
20		$H$17	Atlanta Store Trucks	1	1	Integer
21		$H$18	Miami Store Trucks	2	2	Integer
22		$H$19	Chicago Store Trucks	3	3	Integer
23		$H$20	Los Angeles Store Trucks	4	1	Integer
24		$J$16	Philadelphia Store Tractor-Trailers	1	1	Integer
25		$J$17	Atlanta Store Tractor-Trailers	1	1	Integer
26		$J$18	Miami Store Tractor-Trailers	0	0	Integer
27		$J$19	Chicago Store Tractor-Trailers	1	1	Integer
28		$J$20	Los Angeles Store Tractor-Trailers	1	3	Integer
29						

Source: Microsoft product screenshots used with permission from Microsoft Corporation

FIGURE D-26   Top portion of the Answer Report

	Cell	Name	Cell Value	Formula	Status	Slack
30	Constraints					
31	**Cell**	**Name**	**Cell Value**	**Formula**	**Status**	**Slack**
32	$H$21	Totals: Trucks	9	$H$21<=$F$5	Not Binding	3
33	$J$21	Totals: Tractor-Trailers	6	$J$21<=$F$6	Binding	0
34	$L$16	Philadelphia Store Total Vehicle Capacity	5350	$L$16>=$G$16	Not Binding	338
35	$L$17	Atlanta Store Total Vehicle Capacity	3850	$L$17>=$G$17	Not Binding	337
36	$L$18	Miami Store Total Vehicle Capacity	3000	$L$18>=$G$18	Not Binding	224
37	$L$19	Chicago Store Total Vehicle Capacity	6850	$L$19>=$G$19	Not Binding	220
38	$L$20	Los Angeles Store Total Vehicle Capacity	8550	$L$20>=$G$20	Not Binding	765
39	$H$16	Philadelphia Store Trucks	2	$H$16>=0	Binding	0
40	$H$17	Atlanta Store Trucks	1	$H$17>=0	Binding	0
41	$H$18	Miami Store Trucks	2	$H$18>=0	Binding	0
42	$H$19	Chicago Store Trucks	3	$H$19>=0	Binding	0
43	$H$20	Los Angeles Store Trucks	1	$H$20>=0	Binding	0
44	$J$16	Philadelphia Store Tractor-Trailers	1	$J$16>=0	Binding	0
45	$J$17	Atlanta Store Tractor-Trailers	1	$J$17>=0	Binding	0
46	$J$18	Miami Store Tractor-Trailers	0	$J$18>=0	Binding	0
47	$J$19	Chicago Store Tractor-Trailers	1	$J$19>=0	Binding	0
48	$J$20	Los Angeles Store Tractor-Trailers	3	$J$20>=0	Binding	0
49	$H$16:$H$20=Integer					
50	$J$16:$J$20=Integer					
51						

Source: Microsoft product screenshots used with permission from Microsoft Corporation

FIGURE D-27   Bottom portion of the Answer Report

The Answer Report gives you a wealth of information about the solution. The top portion displays the original and final values of the Objective cell. The second part of the report displays the original and final values of the changing cells. The last part of the report lists the constraints. Binding constraints are those that reached their maximum or minimum value; nonbinding constraints did not.

Perhaps a savings of $724 does not seem significant—however, this problem does not have a specified time frame. The example probably represents one week of shipments for CV Fitness. The store demands will change from week to week, but you could use Solver each time to optimize the truck assignments. In a 50-week business year, the savings that Solver helps you find in shipping costs could be well over $30,000!

Go to the File tab to print the worksheets you created. Save the Excel file as **CV Fitness Trucking Problem.xlsx**, then select the Save As command in the File tab to create a new file called **CV Fitness Trucking Problem 2.xlsx**. You will use the new file in the next section.

EXTENDING THE EXAMPLE

Like all successful companies, CV Fitness looks for ways to grow its business and optimize its costs. Your management team is considering two changes:

- Opening two new stores and expanding the vehicle fleet if necessary
- Improving product design and packaging to reduce the shipping volume of the treadmill from 22 cubic feet to 17 cubic feet

You have been asked to modify your model to see the new requirements for each change separately. The two new stores would be in Denver and Phoenix, and they are 1,040 and 1,470 miles from the Memphis plant, respectively. If necessary, open the CV Fitness Trucking Problem 2.xlsx file, then right-click row 21 at the left worksheet border. Click Insert to enter a new row between rows 20 and 21. Repeat the steps to insert a second new row. Your spreadsheet should look like Figure D-28. Do not worry about the borders for now—you can fix them later.

	A	B	C	D	E	F	G	H	I	J	K	L	M	N
13		Calculations and Results Section:												Cost
14		Distance/Demand Table			Store Demand				Vehicle Loading					
15		Distance Table (from Memphis Plant)	Miles	EB	CT	TM	Volume Required	Trucks	Volume for Trucks	Tractor-Trailers	Volume for Tractor-Trailers	Total Vehicle Capacity	% of Vehicle Capacity Utilized	Shipping Cost
16		Philadelphia Store	1010	140	96	86	5012	2	3000	1	2350	5350	94%	$3,333.00
17		Atlanta Store	380	76	81	63	3513	1	1500	1	2350	3850	91%	$874.00
18		Miami Store	1000	56	64	52	2776	2	3000	0	0	3000	93%	$2,000.00
19		Chicago Store	540	115	130	150	6630	3	4500	1	2350	6850	97%	$2,322.00
20		Los Angeles Store	1810	150	135	180	7785	1	1500	3	7050	8550	91%	$8,869.00
21														
22														
23						Totals:	25716	9	13500	6	14100	27600	93%	$17,398.00
24		Fill Legend:			Changing Cells									Total Cost
25					Optimization Cell									

Source: Microsoft product screenshots used with permission from Microsoft Corporation

FIGURE D-28   Distance/Demand table with two blank rows inserted for the new stores

Enter the two new stores in cells B21 and B22, enter their distances in cells C21 and C22, and enter the Store Demands in cells D21 through F22, as shown in Figure D-29. When you complete this part of the table, fix the borders to include the two new stores. Select the area in the table you want to fix, click the No Borders button to clear the old borders, highlight the area to which you want to add the border, and then click the Outside Borders button.

	A	B	C	D	E	F
13		Calculations and Results Section:				
14		Distance/Demand Table			Store Demand	
15		Distance Table (from Memphis Plant)	Miles	EB	CT	TM
16		Philadelphia Store	1010	140	96	86
17		Atlanta Store	380	76	81	63
18		Miami Store	1000	56	64	52
19		Chicago Store	540	115	130	150
20		Los Angeles Store	1810	150	135	180
21		Denver Store	1040	74	67	43
22		Phoenix Store	1470	41	28	37
23						Totals:
24		Fill Legend:			Changing Cells	
25					Optimization Cell	

Source: Microsoft product screenshots used with permission from Microsoft Corporation

FIGURE D-29   Distance/Demand table with new store locations and demands entered

Next, copy the formulas from cells G20 to N20 to the two new rows in the Vehicle Loading and Cost sections of the table. Select cells G20 to N20, right-click, and click Copy on the menu. Then select cells G21 to N22 and click Paste in the Clipboard group. Your table should look like Figure D-30.

	G	H	I	J	K	L	M	N
13								
14	Vehicle Loading							Cost
15	Volume Required	Trucks	Volume for Trucks	Tractor-Trailers	Volume for Tractor-Trailers	Total Vehicle Capacity	% of Vehicle Capacity Utilized	Shipping Cost
16	5012	2	3000	1	2350	5350	94%	$3,333.00
17	3513	1	1500	1	2350	3850	91%	$874.00
18	2776	2	3000	0	0	3000	93%	$2,000.00
19	6630	3	4500	1	2350	6850	97%	$2,322.00
20	7785	1	1500	3	7050	8550	91%	$8,869.00
21	2839	1	1500	3	7050	8550	33%	$5,096.00
22	1726	1	1500	3	7050	8550	20%	$7,203.00
23	25716	9	13500	6	14100	27600	93%	$17,398.00
24								Total Cost

Source: Microsoft product screenshots used with permission from Microsoft Corporation

FIGURE D-30   Formulas from row 20 copied into rows 21 and 22

Note that most cells in the Totals row have not changed—their formulas need to be updated to include the values in rows 21 and 22. To quickly check which cells you need to update, display the formulas in the Totals row. Hold down the Ctrl key and press the ~ key (on most keyboards, this key is next to the "1" key). The Vehicle Loading and Cost sections now display formulas in the cells (see Figure D-31).

	G	H	I	J	K	L	M	N
13								
14			Vehicle Loading					Cost
15	Volume Required	Trucks	Volume for Trucks	Tractor-Trailers	Volume for Tractor-Trailers	Total Vehicle Capacity	% of Vehicle Capacity Utilized	Shipping Cost
16	=D16*$C$8+E16*$C$9+F16*$C$10	2	=H16*$C$5	1	=J16*$C$6	=I16+K16	=G16/L16	=H16*C16*$D$5+J16*C16*$D$6
17	=D17*$C$8+E17*$C$9+F17*$C$10	1	=H17*$C$5	1	=J17*$C$6	=I17+K17	=G17/L17	=H17*C17*$D$5+J17*C17*$D$6
18	=D18*$C$8+E18*$C$9+F18*$C$10	2	=H18*$C$5	0	=J18*$C$6	=I18+K18	=G18/L18	=H18*C18*$D$5+J18*C18*$D$6
19	=D19*$C$8+E19*$C$9+F19*$C$10	3	=H19*$C$5	1	=J19*$C$6	=I19+K19	=G19/L19	=H19*C19*$D$5+J19*C19*$D$6
20	=D20*$C$8+E20*$C$9+F20*$C$10	1	=H20*$C$5	3	=J20*$C$6	=I20+K20	=G20/L20	=H20*C20*$D$5+J20*C20*$D$6
21	=D21*$C$8+E21*$C$9+F21*$C$10	1	=H21*$C$5	3	=J21*$C$6	=I21+K21	=G21/L21	=H21*C21*$D$5+J21*C21*$D$6
22	=D22*$C$8+E22*$C$9+F22*$C$10	1	=H22*$C$5	3	=J22*$C$6	=I22+K22	=G22/L22	=H22*C22*$D$5+J22*C22*$D$6
23	=SUM(G16:G20)	=SUM(H16:H20)	=SUM(I16:I20)	=SUM(J16:J20)	=SUM(K16:K20)	=SUM(L16:L20)	=G23/L23	=SUM(N16:N20)
24								Total Cost

Source: Microsoft product screenshots used with permission from Microsoft Corporation

FIGURE D-31   Vehicle Loading and Cost sections with formulas displayed in the cells

You must update any Totals cells that do not include the contents of rows 21 and 22. For example, you need to update the Totals cells G23 through L23 and cell N23. Cell M23 is not really a total; it is a cumulative ratio formula, so you do not need to update the cell. Use the following formulas to revise the Totals cells:

- Cell G23: =SUM(G16:G22)
- Cell H23: =SUM(H16:H22)
- Cell I23: =SUM(I16:I22)
- Cell J23: =SUM(J16:J22)
- Cell K23: =SUM(K16:K22)
- Cell L23: =SUM(L16:L22)
- Cell N23: =SUM(N16:N22)

The updated sections should look like Figure D-32.

	G	H	I	J	K	L	M	N
14			Vehicle Loading					Cost
15	Volume Required	Trucks	Volume for Trucks	Tractor-Trailers	Volume for Tractor-Trailers	Total Vehicle Capacity	% of Vehicle Capacity Utilized	Shipping Cost
16	5012	2	3000	1	2350	5350	94%	$3,333.00
17	3513	1	1500	1	2350	3850	91%	$874.00
18	2776	2	3000	0	0	3000	93%	$2,000.00
19	6630	3	4500	1	2350	6850	97%	$2,322.00
20	7785	1	1500	3	7050	8550	91%	$8,869.00
21	2839	1	1500	3	7050	8550	33%	$5,096.00
22	1726	1	1500	3	7050	8550	20%	$7,203.00
23	30281	11	16500	12	28200	44700	68%	$29,697.00
24								Total Cost

Source: Microsoft product screenshots used with permission from Microsoft Corporation

FIGURE D-32   Vehicle Loading and Cost sections with the formulas updated

You are ready to use Solver to determine the optimal vehicle assignment. Click Solver in the Analysis group of the Data tab. You should notice immediately that you must revise the changing cells to include the two new stores; you must also change some of the constraints and add others. Solver has already updated the Objective cell from $N$21 to $N$23 and has updated the $H$23<=$F$5 and $J$23<=$F$6 constraints for vehicle fleet size. To update the changing cells, click the button to the right of the By Changing Variable Cells field and select the cells again, or edit the formula in the window by changing cell address $H$20 to $H$22 and cell address $J$20 to $J$22.

To change a constraint, select the one you want to change, and then click Change (see Figure D-33).

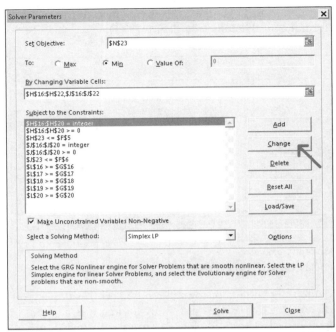

Source: Microsoft product screenshots used with permission from Microsoft Corporation

FIGURE D-33 Selecting a constraint to change

When you click Change, the Change Constraint window appears. Click the Cell Reference button; the selected cells will appear on the spreadsheet with a moving marquee around them (see Figure D-34). Highlight the new group of cells; when the new range appears in the Cell Reference field, click OK. The Solver Parameters window appears with the constraint changed.

	Vehicle Loading						Cost
Volume Required	Trucks	Volume for Trucks	Tractor-Trailers	Volume for Tractor-Trailers	Total Vehicle Capacity	% of Vehicle Capacity Utilized	Shipping Cost
5012	2	3000	1	2350	5350	94%	$3,333.00
3513	1	1500	1	2350	3850	91%	$874.00
2776	2	3000	0	0	3000	93%	$2,000.00
6630	3	4500	1	2350	6850	97%	$2,322.00
7785	1	1500	3	7050	8550	91%	$8,869.00
2839	1	1500	3	7050	8550	33%	$5,096.00
1726	1	1500	3	7050	8550	20%	$7,203.00
30281	11	16500	12	28200	44700	68%	$29,697.00
							Total Cost

Source: Microsoft product screenshots used with permission from Microsoft Corporation

FIGURE D-34 Adding cells H21 and H22 to the Trucks constraint cell range

You also need to update or add the following constraints:

- Update $J$16:$J$20 >=0 to $J$16:$J$22 >=0.
- Update $H$16:$H$20 = integer to $H$16:$H$22 = integer. When changing integer constraints, you must click "int" in the middle field of the Change Constraint window; otherwise, you will receive an error message.
- Update $J$16:$J$20 = integer to $J$16:$J$22 = integer.
- Add constraint $L$21 >= $G$21 (see Figure D-35).
- Add constraint $L$22 >= $G$22.

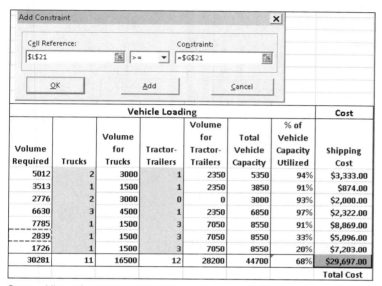

Vehicle Loading							Cost
Volume Required	Trucks	Volume for Trucks	Tractor-Trailers	Volume for Tractor-Trailers	Total Vehicle Capacity	% of Vehicle Capacity Utilized	Shipping Cost
5012	2	3000	1	2350	5350	94%	$3,333.00
3513	1	1500	1	2350	3850	91%	$874.00
2776	2	3000	0	0	3000	93%	$2,000.00
6630	3	4500	1	2350	6850	97%	$2,322.00
7785	1	1500	3	7050	8550	91%	$8,869.00
2839	1	1500	3	7050	8550	33%	$5,096.00
1726	1	1500	3	7050	8550	20%	$7,203.00
30281	11	16500	12	28200	44700	68%	$29,697.00
							Total Cost

FIGURE D-35   Adding a constraint using the Add Constraint window

You are ready to solve the shipping problem to include the new stores in Denver and Phoenix. Figure D-36 shows the updated Solver Parameters window.

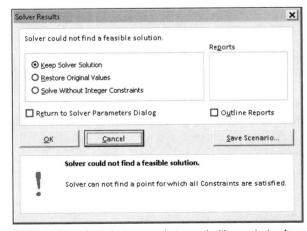

Source: Microsoft product screenshots used with permission from Microsoft Corporation

FIGURE D-36   Solver parameters updated for shipping to seven stores

Before you run Solver again, you might want to attempt to assign the vehicles manually, because your fleet may not be large enough to handle two more stores. In this case, you will quickly realize that the vehicle fleet is at least one truck or tractor-trailer short of the minimum required to ship the needed volume. You can confirm this by running Solver (see Figure D-37).

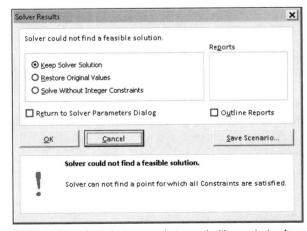

Source: Microsoft product screenshots used with permission from Microsoft Corporation

FIGURE D-37   Vehicle fleet does not meet minimum requirements

The Solver Results window confirms that your truck fleet is too small, so change the value in cell F5 from 12 to 13 to add another truck to your fleet, and then run Solver again. As you add more stores and vehicles to make the problem more complex, Solver will take longer to run, especially on older computers. You may have to wait a minute or more for Solver to finish its iterations and find an answer (see Figure D-38). In this example, Solver recommends that you use 13 trucks and 6 tractor-trailers.

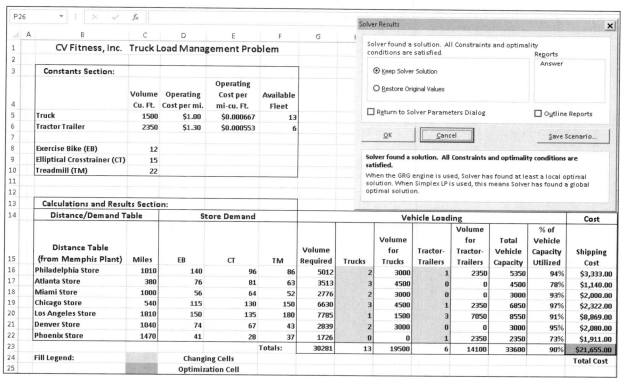

Source: Microsoft product screenshots used with permission from Microsoft Corporation

FIGURE D-38 Solver's solution

Select Answer in the Reports list to add an Answer Report to the workbook, and then click OK. You can keep or delete the old Answer Report 1 tab from the earlier workbook. The new Answer Report is in a new worksheet named Answer Report 2.

You can meet the shipping requirements by adding one more truck, but is it really the most cost-effective solution? What if you add a tractor-trailer instead? Set the number of trucks back to 12, and add a tractor-trailer by entering 7 instead of 6 in cell F6. Run Solver again.

This time Solver finds a less expensive solution, as shown in Figures D-39 and D-40. At first it does not make sense—how can adding a more expensive vehicle (a tractor-trailer) reduce the overall expense? In fact, the additional tractor-trailer has replaced 2 trucks. With 7 tractor-trailers, you only need 11 trucks instead of the original 13.

	Cell	Name	Original Value	Final Value	Integer

Worksheet: [CV Fitness Trucking Problem--Working Copy-12th Edition.xlsx]Problem 2 Truck Mix Changed

Report Created: 7/3/2014 2:12:45 PM

Result: Solver found a solution. All Constraints and optimality conditions are satisfied.

Solver Engine

Engine: Simplex LP

Solution Time: 15.646 Seconds.

Iterations: 6 Subproblems: 15498

Solver Options

Max Time 100 sec, Iterations 100, Precision 0.000001

Max Subproblems Unlimited, Max Integer Sols Unlimited, Integer Tolerance 0%, Assume NonNegative

Objective Cell (Min)

Cell	Name	Original Value	Final Value
$N$23	Totals: Shipping Cost	$21,655.00	$21,389.00

Variable Cells

Cell	Name	Original Value	Final Value	Integer
$H$16	Philadelphia Store Trucks	2	2	Integer
$H$17	Atlanta Store Trucks	3	1	Integer
$H$18	Miami Store Trucks	2	2	Integer
$H$19	Chicago Store Trucks	3	3	Integer
$H$20	Los Angeles Store Trucks	1	1	Integer
$H$21	Denver Store Trucks	2	2	Integer
$H$22	Phoenix Store Trucks	0	0	Integer
$J$16	Philadelphia Store Tractor-Trailers	1	1	Integer
$J$17	Atlanta Store Tractor-Trailers	0	1	Integer
$J$18	Miami Store Tractor-Trailers	0	0	Integer
$J$19	Chicago Store Tractor-Trailers	1	1	Integer
$J$20	Los Angeles Store Tractor-Trailers	3	3	Integer
$J$21	Denver Store Tractor-Trailers	0	0	Integer
$J$22	Phoenix Store Tractor-Trailers	1	1	Integer

Source: Microsoft product screenshots used with permission from Microsoft Corporation

FIGURE D-39   Answer Report 3 displays a more cost-effective solution

		Vehicle Loading					Cost
Volume Required	Trucks	Volume for Trucks	Tractor-Trailers	Volume for Tractor-Trailers	Total Vehicle Capacity	% of Vehicle Capacity Utilized	Shipping Cost
5012	2	3000	1	2350	5350	94%	$3,333.00
3513	1	1500	1	2350	3850	91%	$874.00
2776	2	3000	0	0	3000	93%	$2,000.00
6630	3	4500	1	2350	6850	97%	$2,322.00
7785	1	1500	3	7050	8550	91%	$8,869.00
2839	2	3000	0	0	3000	95%	$2,080.00
1726	0	0	1	2350	2350	73%	$1,911.00
30281	11	16500	7	16450	32950	92%	$21,389.00
							Total Cost

Source: Microsoft product screenshots used with permission from Microsoft Corporation

FIGURE D-40   Seven tractor-trailers and 11 trucks are the optimal mix

You have a solution for the expansion to seven stores. Save your workbook, and then create a new workbook using the Save As command. Name the new workbook **CV Fitness Trucking Problem 3.xlsx**.

Next, evaluate the potential cost savings if the company redesigns its treadmill product and packaging to reduce the shipping volume from 22 cubic feet to 17 cubic feet. Your engineers report that the redesign will cost approximately $10,000. If you can save at least $500 per shipment, the project will pay for itself in less than six months (20 weekly shipments).

Go to cell C10 on the worksheet, replace 22 with 17, and run Solver again. When Solver finds the solution, select Answer to create another Answer Report, and then click OK. See Figure D-41.

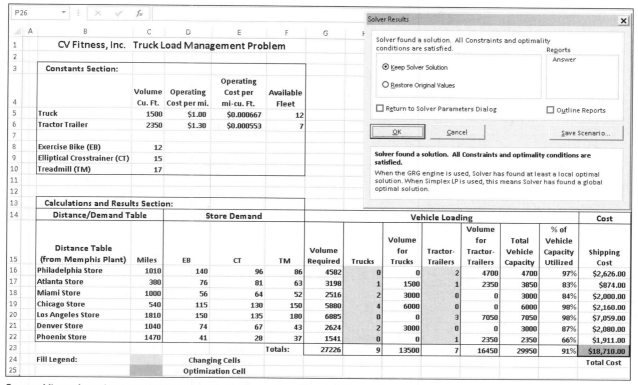

Source: Microsoft product screenshots used with permission from Microsoft Corporation

FIGURE D-41 Solver solution with redesigned treadmill and packaging

Check the Answer Report to see the cost difference between shipping the old treadmills and the redesigned models (see Figure D-42). The cost savings for one shipment is $2,679, which is more than five times the minimum savings you needed. You should go ahead with the project.

	A	B	C	D	E	F
1	**Microsoft Excel 15.0 Answer Report**					
2	**Worksheet: [CV Fitness Trucking Problem--Working Copy-12th Edition.xlsx]Problem 3 Treadmill Repackage**					
3	**Report Created: 7/3/2014 2:29:19 PM**					
4	**Result: Solver found a solution. All Constraints and optimality conditions are satisfied.**					
5	**Solver Engine**					
6	Engine: Simplex LP					
7	Solution Time: 9.969 Seconds.					
8	Iterations: 8 Subproblems: 10238					
9	**Solver Options**					
10	Max Time 100 sec, Iterations 100, Precision 0.000001					
11	Max Subproblems Unlimited, Max Integer Sols Unlimited, Integer Tolerance 0%, Assume NonNegative					
12						
13	Objective Cell (Min)					
14	**Cell**		**Name**	**Original Value**	**Final Value**	
15	$N$23		Totals: Shipping Cost	$21,389.00	$18,710.00	
16						
17	Variable Cells					
18	**Cell**		**Name**	**Original Value**	**Final Value**	**Integer**
19	$H$16		Philadelphia Store Trucks	2	0	Integer
20	$H$17		Atlanta Store Trucks	1	1	Integer
21	$H$18		Miami Store Trucks	2	2	Integer
22	$H$19		Chicago Store Trucks	3	4	Integer
23	$H$20		Los Angeles Store Trucks	1	0	Integer
24	$H$21		Denver Store Trucks	2	2	Integer
25	$H$22		Phoenix Store Trucks	0	0	Integer
26	$J$16		Philadelphia Store Tractor-Trailers	1	2	Integer
27	$J$17		Atlanta Store Tractor-Trailers	1	1	Integer
28	$J$18		Miami Store Tractor-Trailers	0	0	Integer
29	$J$19		Chicago Store Tractor-Trailers	1	0	Integer
30	$J$20		Los Angeles Store Tractor-Trailers	3	3	Integer
31	$J$21		Denver Store Tractor-Trailers	0	0	Integer
32	$J$22		Phoenix Store Tractor-Trailers	1	1	Integer

Source: Microsoft product screenshots used with permission from Microsoft Corporation

FIGURE D-42   Answer Report for the treadmill redesign

When you finish examining the Answer Report, save your file and then close it. To close the workbook, click the File tab and then click Close (see Figure D-43).

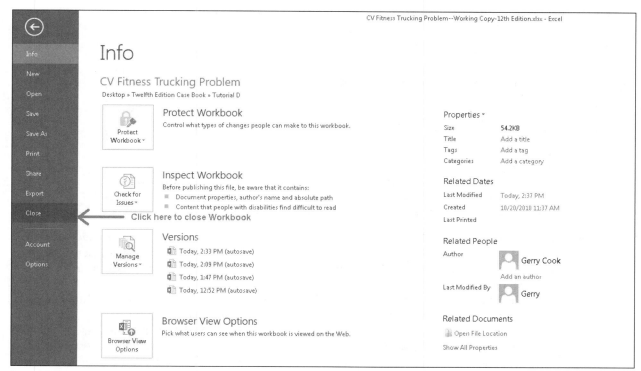

Source: Microsoft product screenshots used with permission from Microsoft Corporation

FIGURE D-43   Closing the Excel workbook

USING SOLVER ON A NEW PROBLEM

A common problem in manufacturing businesses is deciding on a product mix for different items in the same product family. Sensuous Scents Inc. makes a premium collection of perfume, cologne, and body spray for sale in large department stores and boutiques. The primary ingredient is ambergris, a valuable digestive excretion from whales that is harvested without harming the animals. Ambergris costs more than $9,000 per pound and is very difficult to obtain in large quantities; Sensuous Scents can obtain only about 20 pounds of ambergris each year. The other ingredients—deionized water, ethanol, and various additives—are available in unlimited quantities for a reasonable cost.

You have been asked to create a spreadsheet model for Solver to determine the optimal product mix that maximizes Sensuous Scents' net income after taxes.

Setting Up the Spreadsheet

The sections in this spreadsheet are different from those in the preceding trucking problem. You will create a Constants section, a Bill of Materials section for the three products, a Quantity Manufactured section that contains the changing cells, a Calculations section (to calculate ambergris usage, manufacturing costs, and sales revenue per product line), and an Income Statement section to determine the net income after taxes, which will be the optimization cell.

AT THE KEYBOARD

Start a new file called **Sensuous Scents Inc.xlsx** and set up the spreadsheet. *You can also download the spreadsheet skeleton if you prefer; it will save you time.* To access this skeleton file, select Tutorial D from your data files, and then select **Sensuous Scents Inc.xlsx**.

Spreadsheet Title and Constants Section

Your spreadsheet title and Constants section should look like Figure D-44. A discussion of the section entries follows the figure.

▲	A	B	C	D	E	F
1		Sensuous Scents Inc. Product Mix				
2						
3		Constants:		Body Spray	Cologne	Perfume
4		Sales Price per bottle		$11.95	$21.00	$53.00
5		Conversion Cost per Unit (Direct Labor plus Manufacturing Overhead)		$2.60	$6.50	$13.00
6		Minimum Sales Demand		60000	25000	12000
7		Income Tax Rate	0.32			
8		Sales, General and Administrative Expenses per Dollar Revenue	0.30			
9		Available Ambergris (lbs)	20			
10		Cost per lb, Deionized Water	$0.50			
11		Cost per lb, Ethanol	$1.00			
12		Cost per lb, other Additives	$182.00			
13		Cost per lb, Ambergris	$9,072.00			

Source: Microsoft product screenshots used with permission from Microsoft Corporation

FIGURE D-44 Spreadsheet title and Constants section for Sensuous Scents Inc.

- Sales Price per bottle—These values are the sales prices for each of the three products.
- Conversion Cost per Unit—These values are the direct labor costs plus the manufacturing overhead costs budgeted per unit manufactured. A conversion cost is often used in industries that manufacture liquid products.
- Minimum Sales Demand—These values reflect the forecast minimum sales demand that you must supply to your customers. These values will be used later as constraints.
- Income Tax Rate—The rate is 32% of your pretax income. No taxes are paid on losses.
- Sales, General and Administrative Expenses per Dollar Revenue—This value is an estimate of the non-manufacturing costs that Sensuous Scents will incur per dollar of sales revenue. These expenses are subtracted from the Gross Profit value in the Income Statement section to obtain Net Income before taxes.
- Available Ambergris (lbs.)—This value is the amount of ambergris that Sensuous Scents obtained this year for production.
- Cost per lb., Deionized Water—This value is the current cost per pound of deionized water.
- Cost per lb., Ethanol—This value is the current cost per pound of ethanol.
- Cost per lb., other Additives—Scent products contain other additives and fixatives to enhance or preserve the fragrance. This value is the cost per pound of the other additives.
- Cost per lb., Ambergris—This value is the current market price per pound of naturally harvested ambergris. Again, no whales are harmed to obtain the ambergris.

The rest of the cells are filled with a gray background to indicate that you will not use their values or formulas. The section is arranged this way to maintain one column per product all the way down the spreadsheet, which will simplify writing the formulas later.

Bill of Materials Section

Your spreadsheet should contain a Bill of Materials section, as shown in Figure D-45. The section entries are explained after the figure. A bill of materials is a list of raw materials and ingredients required to make one unit of a product.

▲	A	B	C	D	E	F
14						
15		Bill of Materials:		Body Spray	Cologne	Perfume
16		Deionized Water (lb)		0.4	0.1	0.05
17		Ethanol (lb)		0.1	0.02	0.01
18		Other Additives (lb)		0.01	0.001	0.0001
19		Ambergris (lb)		0.0001	0.00018	0.00055

Source: Microsoft product screenshots used with permission from Microsoft Corporation

FIGURE D-45 Bill of Materials section

- Deionized Water (lb.)—The amount of deionized water required to make one unit of each product
- Ethanol (lb.)—The amount of ethanol required to make one unit of each product
- Other Additives (lb.)—The amount of other additives required to make one unit of each product
- Ambergris (lb.)—The amount of ambergris required to make one unit of each product

Extremely small quantities of ambergris and other additives are required to make one bottle of each product. Also, each product requires a different amount of ambergris. Check the values to make sure you entered the correct number of decimal places.

Quantity Manufactured (Changing Cells) Section

This model contains a separate Changing Cells section called Quantity Manufactured, as shown in Figure D-46. This section contains the cells that you want Solver to manipulate to achieve the highest net income after taxes.

	A	B		C	D	E	F
20							
21		Quantity Manufactured (Changing Cells)			Body Spray	Cologne	Perfume
22		Units Produced			60000	25000	12000

Source: Microsoft product screenshots used with permission from Microsoft Corporation

FIGURE D-46   Quantity Manufactured (changing cells) section

Cells D22, E22, and F22 are yellow to indicate that Solver will change them to reach an optimal solution. To begin, enter the minimum sales demand in these cells, which will remind you to specify the minimum demand constraints from the Constants section in the Solver Parameters window.

Calculations Section

Your model should contain the Calculations section shown in Figure D-47.

	A	B	C	D	E	F	G
23							
24		Calculations:		Body Spray	Cologne	Perfume	Totals
25		Lbs of Ambergris Used					
26		Manufacturing Cost per Unit (Materials Costs plus Conversion Cost)					
27		Total Manufacturing Costs per Product Line					
28		Sales Revenues per Product Line					

Source: Microsoft product screenshots used with permission from Microsoft Corporation

FIGURE D-47   Calculations section

The section contains the following calculations:

- Lbs. of Ambergris Used—This value is the pounds of ambergris per unit from the Bill of Materials section, multiplied by Units Produced from the Quantity Manufactured section for each of the three products. The Totals cell (G25) is the sum of cells D25, E25, and F25. Use the value in this cell to specify the constraint that you have only 20 pounds of ambergris available to use for raw materials (Constants section, cell C9).
- Manufacturing Cost per Unit (Materials Costs plus Conversion Cost)—To get this value, write a formula that multiplies the unit cost for each of the four product ingredients by the amount per unit specified in the bill of materials. The total materials costs for the four ingredients are added together, and then the Conversion Cost per Unit is added from the Constants section to obtain the Manufacturing Cost per Unit. Enter the following formula for the Body Spray Manufacturing Cost per Unit in cell D26:

      =$C$10*D16+$C$11*D17+$C$12*D18+$C$13*D19+D5

  Use absolute cell references for the cells that hold values for costs per pound ($C$10, $C$11, $C$12, and $C$13). By doing so, you can copy the body spray formula to the Manufacturing Cost per Unit cells for the cologne and perfume values (cells E26 and F26). The Totals cell (G26) is not used in this row—you can fill the cell in gray to indicate that it is not used.
- Total Manufacturing Costs per Product Line—This value is the Manufacturing Cost per Unit multiplied by Units Produced from the Quantity Manufactured section. The Totals cell (G27) is the sum of cells D27, E27, and F27. You will use the value in the Totals cell in the Income Statement section.

- Sales Revenues per Product Line—This value is the Sales Price per bottle from the Constants section multiplied by Units Produced from the Quantity Manufactured section. The Totals cell (G28) is the sum of cells D28, E28, and F28. You will use the value in this cell in the Income Statement section.

Income Statement Section

The last section you need to construct is the Income Statement, as shown in Figure D-48. An explanation of the needed formulas follows the figure.

Source: Microsoft product screenshots used with permission from Microsoft Corporation

FIGURE D-48   Income Statement section with fill legend

- Sales Revenues—This value is the total sales revenues from the Calculations section (cell G28).
- Less: Manufacturing Cost—This value is the total manufacturing costs from the Calculations section (cell G27).
- Gross Profit—This value is the Sales Revenues minus the Manufacturing Cost.
- Less: Sales, General, and Administrative Expenses—This value is the Sales Revenues multiplied by the Sales, General, and Administrative Expenses per Dollar Revenue from the Constants section (cell C8).
- Net Income before taxes—This value is the Gross Profit minus the Sales, General, and Administrative Expenses.
- Less: Income Tax Expense—If the Net Income before taxes is greater than zero, this value is the Net Income before taxes multiplied by the Income Tax Rate in the Constants section. If Net Income before taxes is zero or less, the Income Tax Expense is zero.
- Net Income after taxes—This value is the Net Income before taxes minus the Income Tax Expense. You will use this value as your optimization cell because you want to maximize Net Income after taxes.

Setting Up Solver

You need to satisfy the following conditions when running Solver:

- Your objective is to maximize Net Income after taxes (cell C37).
- Your changing cells are the Units Produced (cells D22, E22, and F22).
- Observe the following constraints:

  - You must produce at least the Minimum Sales Demand for each product (cells D6, E6, and F6).
  - Your total Lbs. of Ambergris Used (cell G25) cannot exceed the Available Ambergris (cell C9).
  - You cannot produce negative units of any product (enter constraints for the changing cells to be greater than or equal to zero).
  - You can produce only whole units of any product (enter constraints for the changing cells to be integers).

Run Solver and create an Answer Report when Solver finds the solution. When you complete the program, print your spreadsheet with the Solver solution, and print the Answer Report. Save your work and close Excel.

TROUBLESHOOTING SOLVER

Solver is a fairly complex software program. This section helps you address common problems you may encounter when attempting to run Solver.

Using Whole Numbers in Changing Cells

Before you run your first Solver model or rerun a previous model, always enter a positive whole number in each of the changing cells. If you have not already defined maximum and minimum constraints for the values in the changing cells, enter 1 in each cell before running Solver.

Getting Negative or Fractional Answers

If you receive negative or fractional answers when running Solver, you may have neglected to specify one or more of the changing cells as non-negative integers. Alternatively, if you are working on a cost minimization problem and you fail to specify the optimization cell as non-negative, you may receive a negative answer for the cost. Sometimes Solver will also warn you that you have one or more unbounded constraints (see Figure D-49).

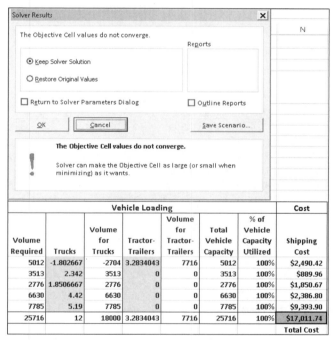

Vehicle Loading							Cost
Volume Required	Trucks	Volume for Trucks	Tractor-Trailers	Volume for Tractor-Trailers	Total Vehicle Capacity	% of Vehicle Capacity Utilized	Shipping Cost
5012	-1.802667	-2704	3.2834043	7716	5012	100%	$2,490.42
3513	2.342	3513	0	0	3513	100%	$889.96
2776	1.8506667	2776	0	0	2776	100%	$1,850.67
6630	4.42	6630	0	0	6630	100%	$2,386.80
7785	5.19	7785	0	0	7785	100%	$9,393.90
25716	12	18000	3.2834043	7716	25716	100%	$17,011.74
							Total Cost

Source: Microsoft product screenshots used with permission from Microsoft Corporation

FIGURE D-49   Solver has an "unbounded" objective function because you did not specify non-negative integer constraints

Creating Overconstrained Models

If Solver cannot find a solution because it cannot meet the constraints you defined, you will receive an error message. When this happens, Solver may even violate the integer constraints you defined in an attempt to find an answer, as shown in Figure D-50. To create this error, the number of trucks available was reduced from 12 to 10.

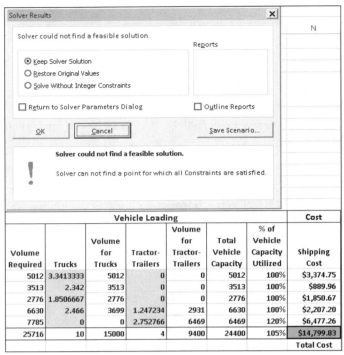

Vehicle Loading							Cost
Volume Required	Trucks	Volume for Trucks	Tractor-Trailers	Volume for Tractor-Trailers	Total Vehicle Capacity	% of Vehicle Capacity Utilized	Shipping Cost
5012	3.3413333	5012	0	0	5012	100%	$3,374.75
3513	2.342	3513	0	0	3513	100%	$889.96
2776	1.8506667	2776	0	0	2776	100%	$1,850.67
6630	2.466	3699	1.247234	2931	6630	100%	$2,207.20
7785	0	0	2.752766	6469	6469	120%	$6,477.26
25716	10	15000	4	9400	24400	105%	$14,799.83
							Total Cost

Source: Microsoft product screenshots used with permission from Microsoft Corporation

FIGURE D-50   Solver could not find a feasible solution because not enough vehicles were available

Setting a Constraint to a Single Amount

Sometimes you may want to enter an exact amount into a constraint, as opposed to a number in a range. For example, if you wanted to assign exactly 11 trucks in the CV Fitness problem instead of a maximum of 12, you would select the equals (=) operator in the Change Constraint window, as shown in Figure D-51.

Source: Microsoft product screenshots used with permission from Microsoft Corporation

FIGURE D-51   Constraining a value to a specific amount

Setting Changing Cells to Integers

Throughout the tutorial, you were directed to set the changing cells to integers in the Solver constraints. In many business situations, there is a logical reason for demanding integer solutions, but this approach does have disadvantages. Forcing integers can sometimes increase the amount of time Solver needs to find a feasible solution. In addition, Solver sometimes can find a solution using real numbers in the changing cells instead of integers. If Solver cannot find a feasible solution or reports that it has reached its calculation time limit, consider removing the integer constraints from the changing cells and rerunning Solver to see if it finds an optimal solution that makes sense.

Restarting Solver with New Constraints

Suppose you want to start over with a completely new set of constraints. In the Solver Parameters window, click Reset All. You will be asked to confirm that you want to reset all the Solver options and cell selections (see Figure D-52).

Source: Microsoft product screenshots used with permission from Microsoft Corporation

FIGURE D-52   Reset options warning

If you want to clear all the Solver settings, click OK. An empty Solver Parameters window appears with all the former entries deleted, as shown in Figure D-53. You can then set up a new model.

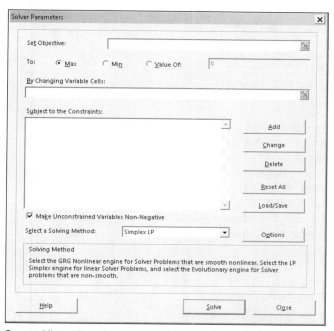

Source: Microsoft product screenshots used with permission from Microsoft Corporation

FIGURE D-53   Solver Parameters window after selecting Reset All

NOTE

Be certain that you want to select the Reset All option before you use it. If you only want to edit, delete, or add a constraint, use the Add, Delete, or Change button for that constraint.

Using the Solver Options Window

Solver has several internal settings that govern its search for an optimal answer. Click the Options button in the Solver Parameters window to see the default selections for these settings, as shown in Figure D-54.

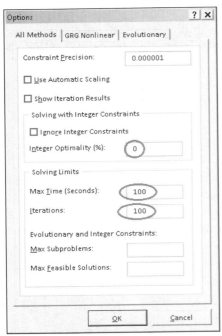

Source: Microsoft product screenshots used with permission from Microsoft Corporation

FIGURE D-54   Solver Options window

You should not need to change the settings in the Options window except for the default value of 5% for Integer Optimality. When it is set at 5%, Solver will get within 5% of the optimal answer, but this setting might not give you the lowest cost or highest income. Change the setting to 0 and click OK.

In more complex problems that have a dozen or more constraints, Solver may not find the optimal solution within the default 100 seconds or 100 iterations. If so, a window will prompt you to continue or stop (see Figure D-55). If you have time, click Continue and let Solver keep working toward the best possible solution. If Solver works for several minutes and still does not find the optimal solution, you can stop by pressing the Ctrl and Break keys together. Click Stop in the resulting window.

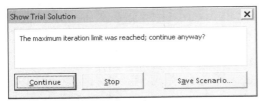

Source: Microsoft product screenshots used with permission from Microsoft Corporation

FIGURE D-55   Prompt that appears when Solver reaches its maximum iteration limit

If you think that Solver needs more time and iterations to reach an optimal solution, you can increase the Max Time and Iterations, but you should probably keep both values under 32,000.

Printing Cell Formulas in Excel

Earlier in the tutorial, you learned how to display cell formulas in your spreadsheet cells. Hold down the Ctrl key and then press the ~ key (on most keyboards, this key is next to the "1" key). You can change the cell widths to see the entire formula by clicking and dragging the column by the dividing lines between the column letters. See Figure D-56.

	Vehicle Loading						Cost
Volume Required	Trucks	Volume for Trucks	Tractor-Trailers	Volume for Tractor-Trailers	Total Vehicle Capacity	% of Vehicle Capacity Utilized	Shipping Cost
=D16*$C$8+E16*$C$9+F16*$C$10	1	=H16*$C$5	1	=J16*$C$6	=I16+K16	=G16/L16	=H16*C16*$D$5+J16*C16*$D$6
=D17*$C$8+E17*$C$9+F17*$C$10	1	=H17*$C$5	1	=J17*$C$6	=I17+K17	=G17/L17	=H17*C17*$D$5+J17*C17*$D$6
=D18*$C$8+E18*$C$9+F18*$C$10	1	=H18*$C$5	1	=J18*$C$6	=I18+K18	=G18/L18	=H18*C18*$D$5+J18*C18*$D$6
=D19*$C$8+E19*$C$9+F19*$C$10	1	=H19*$C$5	1	=J19*$C$6	=I19+K19	=G19/L19	=H19*C19*$D$5+J19*C19*$D$6
=D20*$C$8+E20*$C$9+F20*$C$10	1	=H20*$C$5	1	=J20*$C$6	=I20+K20	=G20/L20	=H20*C20*$D$5+J20*C20*$D$6
=SUM(G16:G20)	=SUM(H16:H20)	=SUM(I16:I20)	=SUM(J16:J20)	=SUM(K16:K20)	=SUM(L16:L20)	=G21/L21	=SUM(N16:N20)

Source: Microsoft product screenshots used with permission from Microsoft Corporation

FIGURE D-56   Spreadsheet with formulas displayed in the cells

To print the formulas, click the File tab and select Print. To restore the screen to its normal appearance and display values instead of formulas, press Ctrl+~ again; the key combination is actually a toggle switch. If you changed the column widths in the formula view, you might have to resize the columns after you change back.

"Fatal" Errors in Solver

When you run Solver, you might sometimes receive a message like the one shown in Figure D-57.

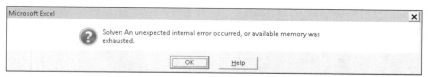

Source: Microsoft product screenshots used with permission from Microsoft Corporation

FIGURE D-57   Fatal error in Solver

Solver usually attempts to find a solution or reports why it cannot. When Solver reports a fatal error, the root cause is difficult to troubleshoot. Possible causes include merged cells on the spreadsheet or printing multiple Answer Reports after running Solver multiple times. A common solution to this error has been to remove the Solver add-in, close Excel, reopen it, and then reinstall Solver. If you encounter a fatal error when using this book, check with your instructor.

Sometimes Solver will generate strange results. Even when your cell formulas and constraints match the ones your instructor has created, Solver's answers might not match the "book" answers. You might have entered your constraints into Solver in a different order, you may have changed some of the options in Solver, or you may have specified real numbers instead of integers for the constraints (or vice versa). Also, the solving method you selected and the amount of time you gave Solver to work can affect the final answer. If your solution is close to the one posted by your instructor, but not exactly the same, show the instructor your setup in the Solver Parameters window. Solver is a powerful tool, but it is not infallible—ask your instructor for guidance if necessary.

SAINT BERNARD BUS LINES

Decision Support Using Microsoft Excel Solver

PREVIEW

Saint Bernard Bus Lines provides passenger transportation services and small-parcel logistics from its headquarters in Philadelphia to major metropolitan areas in the mid-Atlantic region. The company uses a hub-and-spoke network with Philadelphia at its center to provide services to six other cities: Baltimore, Buffalo, Washington, DC, New York, Norfolk, and Pittsburgh. Considering the tremendous growth the company has seen in recent years, management has determined that the current manual system for allocating vehicles has outlived its usefulness. Furthermore, the company is interested in the possibility of selling unused cargo space to other freight companies.

You have been hired as a new MIS consultant to develop a DSS model for Saint Bernard Bus Lines. Your completed model will be used to assign the bus fleet to its six destinations while minimizing costs. You will also modify the model to calculate how taking on additional cargo will affect the company's profitability and operating costs.

PREPARATION

- Review spreadsheet concepts discussed in class and in your textbook.
- Complete any exercises that your instructor assigns.
- Complete any part of Tutorial D that your instructor assigns, or refer to it as necessary.
- Review the file-saving procedures for Windows programs in Tutorial C.
- Review Tutorial F as necessary.

BACKGROUND

You will use your Excel skills to build a decision model and determine how many of each type of bus should be assigned to the six Saint Bernard Bus Lines destinations. The model requires the following data, which the management team has compiled for you:

- Data for the four different types of buses in the fleet:
  - Passenger capacity
  - Cargo space
  - Operating cost per mile (includes fuel, labor, and overhead)
  - Number of available buses
- Ticket pricing to each destination
- Cargo pricing to each destination
- Distance from Philadelphia to each destination

The Marketing department has also given you information about the passenger and cargo demand for each city:

- Expected daily passenger load
- Expected daily cargo load

To satisfy passenger requirements and cargo demand, your Solver model will assign buses by number and type to each destination city. The model will also calculate daily revenues from both passenger service and cargo, as well as the total daily operating cost. You will use the results of these calculations to create a daily gross profit statement. You will run Solver first to minimize the total operating cost. Next, you will modify the model to examine the effect of accepting additional cargo on the total operating cost and profitability. Finally, you will run the modified model to maximize daily gross profits.

Saint Bernard Bus Lines Fleet

The company's fleet consists of 48 buses divided into four different types. The oldest and most expensive model to operate is the D4500 CT Commuter Coach; there are 21 in the fleet. The company also owns eight D4000 Workhorse Commuter Coaches and seven J4500 buses. These two models are less expensive to operate, and they vary in passenger and cargo capacity. Finally, the company has been making inroads into alternative energy by acquiring 12 D4500 CT Hybrid Commuter Coaches. Although these buses are considerably less expensive to operate, they have substantially less cargo capacity than the other models.

ASSIGNMENT 1: CREATING SPREADSHEET MODELS FOR DECISION SUPPORT

In this assignment, you will create spreadsheets that model the business decision Saint Bernard Bus Lines is seeking. In Assignment 1A, you will create a spreadsheet and attempt to assign the buses manually to minimize the total operating cost. In Assignment 1B, you will copy the spreadsheet to a new worksheet and then set up and run Solver to minimize the total operating cost. In Assignment 1C, you will copy the Solver solution to a new worksheet and modify it to add calculations for taking on additional cargo. You will then rerun Solver to determine how the extra cargo affects the company's total operating cost. In Assignment 1D, you will copy the modified Solver spreadsheet and rerun Solver to maximize daily gross profit.

This section helps you set up each of the following spreadsheet components before entering the cell formulas:

- Constants
- Calculations and Results
- Income Statement

The Calculations and Results section is the heart of the decision model. You will set up columns for travel distances, daily demand, bus assignment by type, bus use, and operating costs. The spreadsheet rows will represent destination cities. The Bus Assignment section will hold the range of changing cells for Solver to manipulate. The total operating cost in this section will serve as your optimization cell for Assignments 1B and 1C, and the daily gross profit will be your optimization cell for Assignment 1D. You will add formulas to the additional cargo cells in the Calculations and Results section for Assignments 1C and 1D.

Assignment 1A: Creating the Spreadsheet for the Base Case

A discussion of each spreadsheet section follows. This information helps you set up each section of the model and learn the logic of the formulas in the spreadsheet. If you choose to enter the data directly, follow the cell structure shown in the figures. You can also download the spreadsheet skeleton if you prefer. To access the base spreadsheet skeleton, select Case 8 from your data files, and then select **SaintBernardBusLines.xlsx**.

Constants Section

First, build the skeleton of your spreadsheet. Set up the spreadsheet title and Constants section as shown in Figure 8-1. An explanation of the column items follows the figure.

	Bus Type	Passenger Capacity	Cargo Capacity (lbs)	Operating Cost per mile	Operating Cost per Passenger-Mile	Available Fleet			Fill Legend		
	Saint Bernard Bus Lines Assignment Problem										
	Constants Section:										
	Bus Data:										
	D4500 CT HYBRID COMMUTER COACH	57	2000	$3.50	$0.061404	12				Changing Cells	
	D4500 CT COMMUTER COACH	49	2750	$5.25	$0.107143	21				Optimization Cell	
	D4000 Workhorse Commuter Coach	49	2750	$4.75	$0.096939	8					
	J4500 Coach	59	2000	$5.25	$0.088983	7					
	Fee Schedule:	Average Ticket Price	Cargo Price/lb								
	Destination										
	Baltimore	$90	$0.21								
	Buffalo	$179	$0.76								
	Washington DC	$110	$0.28								
	New York	$98	$0.19								
	Norfolk	$150	$0.55								
	Pittsburgh	$135	$0.61								

Source: Microsoft product screenshots used with permission from Microsoft Corporation

FIGURE 8-1   Spreadsheet title and Constants section

- **Spreadsheet title**—Enter the spreadsheet title in cell B1 and then merge and center the title across cells B1 through F1.
- **Constants section, Bus Data table**—Enter the column headings shown in cells B5 through G5.
- **Bus Type**—Enter each of the four buses listed in cells B6 through B9.
- **Passenger Capacity**—Enter each of the four passenger capacities listed in cells C6 through C9.
- **Cargo Capacity (lbs)**—Enter each of the four cargo capacities listed in cells D6 through D9.
- **Operating Cost per mile**—Enter each of the four operating costs per mile listed in cells E6 through E9.
- **Operating Cost per Passenger-Mile**—This value is the Operating Cost per mile divided by the passenger capacity of the bus, but you do not need to enter formulas. Enter each of the four operating costs per passenger mile listed in cells F6 through F9. These values are not used in the Solver solution but are provided as an aid to completing Assignment 1A.
- **Available Fleet**—This value is the number of buses of each type that Saint Bernard Bus Lines keeps in service. Enter these numbers in cells G6 through G9.
- **Constants section, Fee Schedule table**—Enter the column headings shown in cells B11 through D11.
- **Destination**—Enter the six destination cities in cells B12 through B17.
- **Average Ticket Price**—Enter the passenger ticket prices for the six destinations in cells C12 through C17.
- **Cargo Price/lb**—Enter the cargo price per pound for the six destinations in cells D12 through D17.
- **Fill Legend**—This section is actually adjacent to the Constants section. Enter "Fill Legend" in cell J5, fill cell J6 in yellow, fill cell J7 in orange, enter "Changing Cells" in cell K6, and enter "Optimization Cell" in cell K7.

Calculations and Results Section

The Calculations and Results section (see Figure 8-2) will contain mileages, daily passenger bookings, and daily cargo shipment data obtained from the Marketing department. Although these values are constants, keeping them in the Calculations and Results section facilitates writing and copying formulas in the bus utilization, costs, and cargo columns. This section also includes the Bus Assignment table, which contains the changing cells and calculations for bus usage, costs, and additional cargo. An explanation of the sections and columns follows the figure.

		Calculations and Results Section:	Daily Demand		Bus Assignment				Bus Utilization				Costs	Extra
19														
20	Destination	Distance from Philadelphia Hub	Daily Passenger Bookings	Daily Cargo Shipments Lbs	D4500 CT HYBRID Assigned	D4500 CT COMMUTER Assigned	D4000 Workhorse Assigned	J4500 Coach Assigned	Total Passenger Capacity	% of Passenger Capacity Utilized	Total Cargo Capacity	% of Cargo Capacity Utilized	Operating Cost	Add'l Cargo Lbs to be Added
21	Baltimore	106	500	7,500	1	1	1	1						
22	Buffalo	380	250	5,000	1	1	1	1						
23	Washington DC	139	550	7,500	1	1	1	1						
24	New York	95	700	7,000	1	1	1	1						
25	Norfolk	277	200	3,500	1	1	1	1						
26	Pittsburgh	304	250	4,500	1	1	1	1						
27		Total/Avg												
28													Total Cost	

Source: Microsoft product screenshots used with permission from Microsoft Corporation

FIGURE 8-2   Calculations and Results section

- **Table headings**—If you did not use the skeleton spreadsheet, enter the column headings shown in cells B19 through O20 in Figure 8-2.
- **Destination**—Cells B21 through B26 hold the six cities serviced daily by Saint Bernard Bus Lines.
- **Distance from Philadelphia Hub**—Cells C21 through C26 hold the route distances in miles to each of the six destinations.
- **Daily Passenger Bookings**—Cells D21 through D26 hold the average number of passenger tickets booked each day.
- **Daily Cargo Shipments Lbs**—Cells E21 through E26 hold the average number of pounds of cargo shipped daily.

- Bus Assignment section—Cells F21 through I26 are the heart of the Solver model—the changing cells. The cells hold the amounts of each of the four bus types that Solver will assign to the six destinations. Enter "1" in each of these cells for now. You should fill the cells with a background color to indicate that they are the changing cells for Solver. To fill the cells, select them and then click the Fill Color button in the Font group on the Home tab. In the spreadsheet skeleton, the cells are yellow.
- Bus Utilization section, Total Passenger Capacity—Cells J21 through J26 hold the total passenger capacity for each destination. The capacity is calculated by multiplying the number of each assigned bus type by its passenger capacity, which is taken from cells C6 through C9 of the Constants section. Next, take the sum of the total capacities for the four types of buses assigned. Be sure to use absolute cell references for the passenger capacity values from the Constants section so that you have to write the formula only for the first cell (J21); then you can copy the formula to cells J22 through J26.
- % of Passenger Capacity Utilized—Cells K21 through K26 hold the percentage of passenger capacity used for each destination. The value is calculated by dividing Daily Passenger Bookings by Total Passenger Capacity.
- Total Cargo Capacity—Cells L21 through L26 hold the total cargo capacity for each destination. The capacity is calculated by multiplying the number of each assigned bus type by its cargo capacity, which is taken from cells D6 through D9 of the Constants section. Next, take the sum of the total capacities for the four types of buses assigned. Again, you must use absolute cell references for the cargo capacity values from the Constants section so that you only have to write the formula for the first cell (L21); then you can copy the formula to the other five cells.
- % of Cargo Capacity Utilized—Cells M21 through M26 hold the percentage of cargo capacity used. The percentage is calculated by dividing the Daily Cargo Shipments by the Total Cargo Capacity.
- Operating Cost—Cells N21 through N26 hold the operating cost for each bus type to each destination. The cost is calculated by the following formula:
  Number of buses assigned × Operating Cost per mile × Mileage to destination
  The costs for each of the four buses are then added to obtain the operating cost. Remember to use absolute cell references for the Operating Cost per mile, which is listed in cells E6 through E9 of the Constants section. That way, you have to write the formula only for the first cell (N21); then you can copy the formula to the other five cells.
- Add'l Cargo Lbs to be Added—Leave cells O21 through O26 blank for now. You will place formulas in these cells in Assignment 1C.
- Total/Avg—Total the cell entries for every column except column K (% of Passenger Capacity Utilized) and column M (% of Cargo Capacity Utilized). Place the totals in cells D27 to J27, L27, and N27. For cell K27, divide the total Daily Passenger Bookings in cell D27 by the Total Passenger Capacity in cell J27 and format the result as a percentage. For cell M27, divide the total Daily Cargo Shipments in cell E27 by the Total Cargo Capacity in cell L27 and format the result as a percentage. These two averages are the overall utilization rates of your fleet. For now, leave cell O27 blank (total for Add'l Cargo Lbs to be Added).

If you wrote your formulas correctly, the Calculations and Results section should look like Figure 8-3.

	A	B	C	D	E	F	G	H	I	J	K	L	M	N	O
18															
19		Calculations and Results Section:			Daily Demand		Bus Assignment				Bus Utilization				Costs	Extra
20		Destination		Distance from Philadelphia Hub	Daily Passenger Bookings	Daily Cargo Shipments Lbs	D4500 CT HYBRID Assigned	D4500 CT COMMUTER Assigned	D4000 Workhorse Assigned	J4500 Coach Assigned	Total Passenger Capacity	% of Passenger Capacity Utilized	Total Cargo Capacity	% of Cargo Capacity Utilized	Operating Cost	Add'l Cargo Lbs to be Added
21		Baltimore		106	500	7,500	1	1	1	1	214	234%	9,500	79%	$ 1,988	
22		Buffalo		380	250	5,000	1	1	1	1	214	117%	9,500	53%	$ 7,125	
23		Washington DC		139	550	7,500	1	1	1	1	214	257%	9,500	79%	$ 2,606	
24		New York		95	700	7,000	1	1	1	1	214	327%	9,500	74%	$ 1,781	
25		Norfolk		277	200	3,500	1	1	1	1	214	93%	9,500	37%	$ 5,194	
26		Pittsburgh		304	250	4,500	1	1	1	1	214	117%	9,500	47%	$ 5,700	
27			Total/Avg	2,450	35,000	6	6	6	6	1,284	191%	57,000	61%	$ 24,394		
28														Total Cost	

Source: Microsoft product screenshots used with permission from Microsoft Corporation

FIGURE 8-3 Completed Calculations and Results section

Income Statement Section

The Income Statement section (see Figure 8-4) is actually a projection of daily gross profits and is based on the number of buses that will be assigned either manually or by Solver. An explanation of the line items follows the figure.

	A	B	C
28			
29		Income Statement Section:	
30		Passenger Revenues:	
31		Cargo Revenues:	
32		Additional Cargo	
33		Total Revenues:	
34		less Operating Costs:	
35		Daily Gross Profit:	

Source: Microsoft product screenshots used with permission from Microsoft Corporation

FIGURE 8-4   Income Statement section

- Passenger Revenues—This value is calculated by multiplying the passenger tickets booked for each destination (cells D21 through D26) by their respective average ticket prices (cells C12 through C17), and then totaling the ticket revenues for the six destinations.
- Cargo Revenues—This value is calculated by multiplying the daily cargo shipments for each destination (cells E21 through E26) by their respective cargo prices (cells D12 through D17), and then totaling the cargo revenues for the six destinations.
- Additional Cargo—This value is the additional revenue from cell O27, which will be used later in the assignment.
- Total Revenues—This value is the total of Passenger Revenues, Cargo Revenues, and Additional Cargo revenues.
- Less Operating Costs—This value is the Total Cost from cell N27.
- Daily Gross Profit—This value is the Total Revenues minus the operating costs. This cell will be used as the optimization cell for Assignment 1D.

If your formulas are correct, the initial Income Statement section will appear as shown in Figure 8-5.

	A	B	C
28			
29		Income Statement Section:	
30		Passenger Revenues:	$ 296,850.00
31		Cargo Revenues:	$ 13,480.00
32		Additional Cargo	
33		Total Revenues:	$ 310,330.00
34		less Operating Costs:	$ 24,393.75
35		Daily Gross Profit:	$ 285,936.25

Source: Microsoft product screenshots used with permission from Microsoft Corporation

FIGURE 8-5   Initial income statement

The initial income statement correctly reflects the revenues expected from the passenger and cargo bookings, but the operating costs are not correct because the buses required to transport the passengers and cargo have not been completely assigned yet.

Attempting a Manual Solution

Attempt to assign your bus fleet manually in the spreadsheet. You have several good reasons for doing this. First, you can make sure your model is working correctly before you set up Solver to run. Second, assigning the bus fleet manually will demonstrate which constraints you must meet in solving the problem. For instance, if a passenger or cargo utilization rate is over 100%, you have not assigned enough buses to carry all the passengers or cargo to a particular destination. Therefore, one constraint is that the total passenger capacity for the buses assigned to a destination must be greater than or equal to the passenger bookings. Another constraint is that the total cargo capacity for the buses assigned must be greater than the cargo shipments booked. Given the fleet size, you can probably assign the fleet manually and meet all of your constraints. However, will your total operating cost be the least expensive solution? Running the problem manually will provide an initial operating cost to which you can compare your Solver solution later. The Solver optimization tool should give you a better solution than assigning the fleet manually.

When attempting to assign the buses manually in the Bus Assignment section (the changing cells), you want to assign the most cost-efficient buses first, which is why the Operating Cost per Passenger-Mile data is included in the Constants section. You will want to assign all the D4500 CT Hybrid Commuter Coach buses first, then the J4500 Coach, then the D4000 Workhorse Commuter Coach buses, and finally the D4500 CT Commuter Coach buses. You must satisfy both the passenger and cargo demands for each destination—in other words, the Total Passenger Capacity values in cells J21 through J26 and the Total Cargo Capacity values in cells L21 through L26 must be equal to or greater than the Daily Demand values in cells D21 through D26 and E21 through E26. If you have satisfied the passenger and cargo demands correctly, none of the utilization rates in cells K21 through K26 and M21 through M26 will exceed 100 percent. In addition, the total buses assigned for each type (cells F27 through I27) cannot exceed the available number of each bus type (cells G6 to G9).

When you reach a solution that satisfies the preceding constraints, save your workbook. Name the worksheet **Saint Bernard Guess** and then right-click the worksheet name tab, click Move or Copy, copy the worksheet, and rename the new copy **Saint Bernard Solver 1**. You will use the new worksheet to complete the next part of the assignment.

Assignment 1B: Setting Up and Running Solver

Before using the Solver Parameters window, you should jot down the parameters you must define and their cell addresses. Here is a suggested list:

- The cell you want to minimize (Total Cost, cell N27)
- The cells you want Solver to manipulate to obtain the optimal solution (Bus Assignment, cells F21 through I26)
- The constraints you must define:
  - All the bus assignment cells must contain non-negative integers.
  - The total number of each type of bus assigned (cells F27 through I27) cannot exceed the number of available buses of each type (cells G6 through G9).
  - The total passenger capacity to each destination (cells J21 through J26) must be equal to or greater than the total passenger bookings for each destination (cells D21 through D26).
  - The total cargo capacity to each destination (cells L21 through L26) must be equal to or greater than the total cargo shipment for each destination (cells E21 through E26).

Next, set up your problem. In the Analysis group on the Data tab, click Solver; the Solver Parameters window appears, as shown in Figure 8-6. Enter "Total Cost" in the Set Objective text box, click the Min button, designate your changing cells (cells F21 through I26), and add the constraints from the preceding list. Use the default Simplex LP solving method. If you need help defining your constraints, refer to Tutorial D.

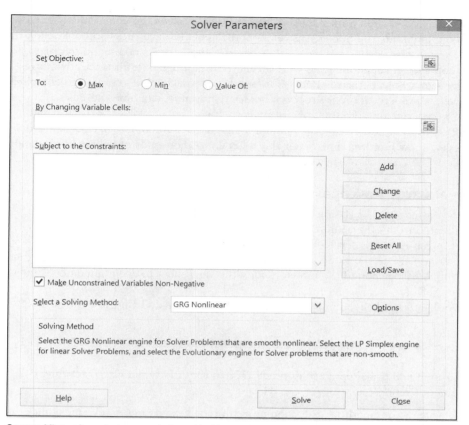

Source: Microsoft product screenshots used with permission from Microsoft Corporation
FIGURE 8-6   The Solver Parameters window

Next, you should click the Options button and check the Options window that appears (see Figure 8-7). The default Integer Optimality is 5%; change it to 1% to get a better answer. Make sure the Constraint Precision is set to the default value of 0.000001 and that the Use Automatic Scaling option is checked. When you finish setting the options, click OK to return to the Solver Parameters window.

Source: Microsoft product screenshots used with permission from Microsoft Corporation
FIGURE 8-7   The Solver Options window

Run Solver and click Answer Report when Solver finds a solution that satisfies the constraints. When you finish, print the entire workbook, including the Solver Answer Report Sheet. To save the workbook, click the File tab and then click Save. For the rest of the case, you either can use the Save As command to create new Excel workbooks or continue copying and renaming the worksheets. Both options offer distinct advantages, but having all of your worksheets and Solver Answer Reports in one Excel workbook allows you to compare different solutions easily, as well as prepare summary reports. Before continuing, examine the bus assignments that Solver chose for minimizing the total operating cost. If you set up Solver correctly, you should see a significant reduction in total cost from your manual assignment. Remember that these are daily values, so the ability to save even a few hundred dollars per day adds up to thousands of dollars by the end of the year.

Assignment 1C: Additional Cargo Revenues

Using the model you created in Assignment 1A, Saint Bernard Bus Lines can also determine the revenue benefit of accepting additional cargo from other companies. Taking on additional cargo might be a good option for Saint Bernard Bus Lines if its cargo service has excess capacity. After contacting other parcel delivery companies in the area that service the same six destinations from Philadelphia, you determine that the rate for additional cargo will be $1 per pound. You must modify the worksheet to accommodate selling the excess capacity and then rerun Solver.

Copy your Saint Bernard Solver 1 worksheet and rename it **Saint Bernard Add'l Cargo**. Perform the following tasks:

- You must insert a calculation for the cells in the Additional Cargo column (cells O21 through O26).
- You must link the total additional cargo sold in cell O27 to the Additional Cargo revenue cell in the Income Statement section (cell C32).

By selling additional cargo space to competitors, the company still receives passenger revenues, but at the same time it collects additional revenues from exploiting excess capacity. You must use an If statement to write the formulas for cells O21 through O26. Use the following logic for the formula you place in cell O21:

- If the Total Cargo Capacity to the Baltimore destination (cell L21) is greater than the Daily Cargo Shipments (cell E21), the company can sell the difference between the two (its excess capacity) to competitors.
- If the Total Cargo Capacity to the Baltimore destination (cell L21) is less than or equal to the Daily Cargo Shipments (cell E21), the company has no excess capacity to sell to competitors.

Click the Insert Function button (fx) next to the formula bar below the Ribbon to help you write the If statement (see Figure 8-8). After you enter the formula in cell O21, you can copy and paste the formula into cells O22 through O26. Cell O27 is the total of cells O21 through O26.

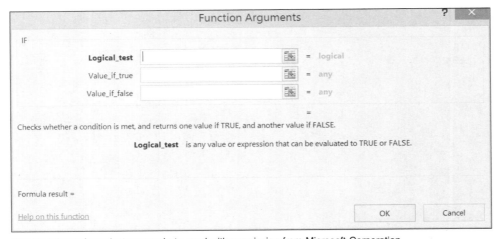

Source: Microsoft product screenshots used with permission from Microsoft Corporation

FIGURE 8-8   The If function window

Next, enter "=O27" in cell C32 of the Income Statement section. Because the revenue for selling excess capacity is $1 per pound, you do not need to enter a formula to multiply the pounds of cargo by the cargo cost per pound. If the cost were less or more than $1, you would have to include the cost in the Constants section and then multiply the pounds sold by the cost per pound to obtain the total excess cargo revenue.

Although you are now taking extra cargo and increasing your revenues, is the current bus assignment still the most cost efficient? Because the excess capacity sold does not affect your costs, you could say that the current bus assignment still works. Instead of minimizing costs, however, what if you were interested in maximizing revenues? In other words, is gross profit greater than that in the earlier solutions? Should you run Solver again to maximize gross profit instead of minimizing operating costs? Because you added the ability to sell excess cargo in the business model, you should probably maximize gross profit to account for any possible efficiencies.

Assignment 1D: Rerunning Solver to Maximize Gross Profit

Copy the worksheet that contains the solution for excess capacity, and rename the new worksheet **Saint Bernard Solver 2**. Click Solver to open the Solver Parameters window, and then change the value in the Set Objective text box to $C$35 (the cell that contains Daily Gross Profit). Click the Max button to maximize the Set Objective value, as shown in Figure 8-9, and then run Solver.

Source: Microsoft product screenshots used with permission from Microsoft Corporation
FIGURE 8-9   Changing the objective and solving for maximized profit

The Solver Results window appears (see Figure 8-10) and warns that the linearity conditions required by the Simplex LP solving method are not satisfied. The default Simplex calculation method will not work because the model now includes the excess cargo equations, which are If functions and therefore nonlinear.

Source: Microsoft product screenshots used with permission from Microsoft Corporation
FIGURE 8-10   Error message when Simplex LP method is used on nonlinear models

Fortunately, Solver has methods for working with nonlinear problems as well. Open the Solver Parameters window again and click the Select a Solving Method list arrow. Click GRG Nonlinear (see Figure 8-11), and then click Options. Under Options, click the All Methods tab to check that the Integer Optimality is set to 1% (see Figure 8-12), and then click OK to return to the Solver Parameters window. Run Solver again; the solution will probably take longer than in the earlier problem. If the solution takes too long to calculate, you can cap the Max Time in seconds in the Options window (see Figure 8-12). Click Answer Report and then click OK to create a second Answer Report. Examine it and the worksheet to see if maximizing daily gross revenues provides a better bus assignment solution when compared with the solution that minimized total costs.

> ☑ Make Unconstrained Variables Non-Negative
>
> Select a Solving Method: GRG Nonlinear   ∨
>
> Solving Method
> Select the GRG Nonlinear engine for Solver Problems that are smooth nonlinear. Select the LP S
> for linear Solver Problems, and select the Evolutionary engine for Solver problems that are non-

Source: Microsoft product screenshots used with permission from Microsoft Corporation

FIGURE 8-11   Changing the solving method

Source: Microsoft product screenshots used with permission from Microsoft Corporation

FIGURE 8-12   Solver Options window with Integer Optimality set to 1%

Print the Saint Bernard Solver 2 worksheet and Answer Report 2, and then save your workbook.

ASSIGNMENT 2: USING THE WORKBOOK FOR DECISION SUPPORT

You have built a series of worksheets to determine the best bus assignments with and without selling excess cargo space. You will now complete the case by using your solutions and Answer Reports to make recommendations in a memorandum. Use Microsoft Word to write a memo to the management team at Saint Bernard Bus Lines. State the results of your analysis, whether you think the current bus assignment method is still profitable, and whether you think bus assignments should be based on lowest cost or maximum profit.

- Set up your memo as described in Tutorial E.
- In the first paragraph, briefly describe the situation and state the purpose of your analysis.

- Next, summarize the results of your analysis and give your recommendations.
- Support your recommendation with appropriate screen shots or Excel objects from the Excel workbook. (Tutorial C describes how to copy and paste Excel objects.)
- In a future case, you might suggest revisiting the Solver analysis to determine the profitability of making changes to the bus fleet.

ASSIGNMENT 3: GIVING AN ORAL PRESENTATION

Your instructor may request that you summarize your analysis and recommendations in an oral presentation. If so, prepare a presentation for the CEO and other managers that lasts 10 minutes or less. When preparing your presentation, use PowerPoint slides or handouts that you think are appropriate. Tutorial F explains how to prepare and give an effective oral presentation.

DELIVERABLES

Prepare the following deliverables for your instructor:

- A printout of the memo
- Printouts of your worksheets and Answer Reports
- Your Word document, Excel workbook, and PowerPoint presentation on electronic media or sent to your course site as directed by your instructor

Staple the printouts together with the memo on top. If you have more than one Excel workbook file for your case, write your instructor a note that describes the different files.

THE FUND OF FUNDS INVESTMENT MIX DECISION

Decision Support Using Microsoft Excel Solver

PREVIEW

You have been hired as a new mutual fund manager and have $100 million to invest. In this case, you will use Excel Solver to determine the best mix of securities to buy.

PREPARATION

- Review spreadsheet concepts discussed in class and in your textbook.
- Complete any exercises that your instructor assigns.
- Complete any part of Tutorial D that your instructor assigns, or refer to it as necessary.
- Review the file-saving procedures for Windows programs in Tutorial C.
- Review Tutorial F as necessary.

BACKGROUND

You have been hired as a new fund manager by a large investment company that runs a number of mutual funds. You have been given $100 million to invest and have been assigned to start a new fund. It is up to you to invest the money wisely.

A mutual fund invests its money in securities, such as stocks and bonds. For example, a mutual fund might have $10 million to invest. Assume that the entire amount is invested in individual shares of common stocks from companies such as General Motors, DuPont, and Bank of America. The mutual fund company's own common stock is also traded on the stock exchange. To continue the example, assume that the mutual fund starts with 1 million shares outstanding. The starting price of a share of its common stock would be $10 million divided by 1 million shares, or $10 a share. Assume further that the value of the purchased common stocks increases and is now worth $11 million. The value of a share of the mutual fund's common stock would now be $11 million divided by 1 million shares, or $11.00.

A mutual fund must have an investment strategy that governs its investment purchases. For example, a fund might invest in corporate bonds or government bonds, or the fund might invest in stocks of well-established industrial companies. A fund might also invest in stocks of companies devoted to real estate development—such funds are usually called REITs (real estate investment trusts).

Instead of buying shares of individual companies, a mutual fund can buy shares *of other mutual funds.* Such a fund is called a "fund of funds." A diversified fund of funds would invest some of its money in funds devoted to bonds, in funds devoted to well-established industrial companies, in funds devoted to REITs, and in other kinds of funds.

You decide to adopt this diversified fund of funds strategy. You will invest in funds oriented to industrial companies, funds for high-grade bonds, REITs, and funds oriented to precious metals, such as gold and silver.

You expect each kind of fund to have different rates of return, different burdens for administrative expenses, and different risk levels. Your decision making will be constrained by certain rules imposed by management. Each of these factors is described next.

Rates of Return

A fund could gain revenue in three ways: (1) from interest earned on bonds and other fixed-income securities, (2) from dividends on common stocks, and (3) from increases in the value of securities owned.

For example, say that shares of XYZ Company sell for $100, and that a fund invests $100,000 in 1,000 shares of XYZ stock. XYZ pays a dividend of $1.00 per share, which means that the fund receives $1,000 from XYZ. Also, the shares increase in value by $10 per share, which is a total of $10,000. The fund would recognize revenue of $1,000 in dividends and $10,000 in increased market value, for a total of $11,000. Comparing this total with the initial value of $100,000 means that the fund's rate of return is $11,000 divided by $100,000, or 11 percent.

As another example, say that a 10-year government bond with a face value of $1,000 pays interest at 4 percent. A fund that invests in these bonds would earn a 4 percent rate of return, assuming that the market value of the bond does not change. Of course, the market value can change, which would influence the overall rate of return.

Note that the market value of securities can also decrease, which means that rates of return can be low or even negative.

If a fund invests in other funds, its rate of return will be the average from the funds in which you invested. Figure 9-1 shows the average rates of returns that you expect from the various investments.

Type of Fund	Expected Average Rate of Return (%)
Stocks of industrial companies	8
High-grade bonds	4
Real estate companies (REITs)	12
Precious metals companies	10

Source: © 2015 Cengage Learning®

FIGURE 9-1 Expected average rates of return

As an example of how this data will be used, assume that your fund of funds invests $10 million in funds that own shares of precious metals companies. The average rate of return for those funds is expected to be 10 percent, which means your investments would create $1 million of revenue from your fund of funds.

Administrative Expense Burden

Your new employer has many analysts who follow the performance of investment securities you might buy. You call on these analysts to suggest funds for purchase, follow the daily progress of the funds, and follow the stocks and bonds that the funds own. These analysts are expensive; they have high salaries as well as office expenses, computer service expenses, and so on. Some kinds of securities are easier to analyze than others, so their administrative costs depend on the kind of fund. Assume that stocks generally require more oversight than bonds, and that some kinds of stocks require more work than others. The expected administrative cost rates for each kind of fund are shown in Figure 9-2.

Type of Fund	Expected Average Administrative Cost Rate
Stocks of industrial companies	.008 (0.8%)
High-grade bonds	.002 (0.2%)
Real estate companies (REITs)	.007 (0.7%)
Precious metals companies	.005 (0.5%)

Source: © 2015 Cengage Learning®

FIGURE 9-2 Expected average administrative cost rates

As an example of how this data would be used, assume that your fund of funds invests $10 million in funds that own shares of precious metals companies. The expected administrative cost rate for the funds you own is expected to be one-half of one percent (0.5%). This means that precious metals investments in your fund of funds would cost $50,000 in administrative expenses.

Risk Levels

The value of a security can decrease. If a security has lost value and then is sold, the owner will lose money. *Risk* is a term that indicates the likelihood of losing money on an investment. Some kinds of investments are riskier than others. For example, stocks are generally thought to be riskier investments than high-grade bonds. Your company's research department has assigned risk factors to the different kinds of funds you want to buy. Factor values range from 1 to 5. A risk factor of 1 represents low risk, and a risk factor of 5 represents high risk. Assigned risk factors are shown in Figure 9-3.

Type of Fund	Risk Factor
Stocks of industrial companies	3
High-grade bonds	1
Real estate companies (REITs)	4
Precious metals companies	5

Source: © 2015 Cengage Learning®

FIGURE 9-3   Assigned risk factors

As an example of how this data would be used, assume that your fund of funds invests $2 million in each type of fund. An average risk factor could be computed for the fund of funds, as shown in Figure 9-4.

Type of Fund	Invested ($)	Risk Factor	Risk Points = Invested * Risk Factor
Stocks of industrial companies	2 million	3	6 million
High-grade bonds	2 million	1	2 million
Real estate companies (REITs)	2 million	4	8 million
Precious metals companies	2 million	5	10 million
		Total	26 million (A)
		Invested	$8 million (B)
		Average risk factor (A/B)	3.25

Source: © 2015 Cengage Learning®

FIGURE 9-4   Calculation of average risk factor for fund of funds

The average risk factor can be used as a way of controlling overall risk in the fund of funds. You might decide to invest conservatively so that the overall risk factor does not exceed 3.5, for example. This kind of rule would prevent you from overinvesting in high-yielding kinds of funds that also carry extensive risk.

Management Rules

Your management said you would have a free hand when building your fund, but that statement is not completely true. You must follow certain rules that your boss says are intended to reduce risk. You have $100 million to invest. You can keep $1 million in cash, which means you must invest at least $99 million. You must invest at least $10 million in each kind of fund, and your average risk factor cannot exceed 3.1. The sum of the amount invested in industrial stock funds and bond funds must be at least 60 percent of the total invested in all types of funds. These rules will constrain your investment decisions.

These rules reflect management's consensus thinking on today's investment environment. However, interest rates on corporate and U.S. government bonds have been very low for a number of years. Some of your managers think that rates will rise soon, and that the risk factor assigned to bonds is too low.

ASSIGNMENT 1: CREATING A SPREADSHEET FOR DECISION SUPPORT

In this assignment, you will produce spreadsheets that model the business decision. In Assignment 1A, you will create a Solver spreadsheet to model the fund of funds investment decision. This spreadsheet will be the base case. In Assignment 1B, the extension case, you will create a Solver spreadsheet to model the investment decision with a different outlook on interest rates.

In Assignments 2 and 3, you will use the spreadsheet models to develop information needed to recommend the best investment mix for your mutual fund. In Assignment 2, you will document your recommendation in a memorandum; in Assignment 3, you will give your recommendations in an oral presentation.

Your spreadsheets for this assignment should include the cells explained in the following sections. You will set up the cells in each of the following spreadsheet sections before entering cell formulas. Your spreadsheets will also include decision constraints, which you will enter using Solver.

- Changing Cells
- Constants
- Calculations
- Income Statement

Assignment 1A: Creating the Spreadsheet for the Base Case

A discussion of each spreadsheet section follows. The discussion explains how to set up each section and explains the logic of the formulas in the section's cells. You can enter the data yourself or use the spreadsheet skeleton to save time. To access the spreadsheet skeleton, select Case 9 in your data files, and then select **FundOfFundsBase.xlsx**.

Changing Cells Section

Your spreadsheet should have the changing cells shown in Figure 9-5.

1	**The Fund of Funds Allocation Decision**		
2	**Changing Cells**		
3	Amount invested in Stock funds	$	1.00
4	Amount invested in Bond funds	$	1.00
5	Amount invested in Precious Metals funds	$	1.00
6	Amount invested in REIT funds	$	1.00

Source: Microsoft product screenshots used with permission from Microsoft Corporation

FIGURE 9-5   Changing Cells section

You will ask the Solver model to compute how much money to invest in each kind of mutual fund. Start with a value of "$1.00" in each cell. Solver will change each $1 value as it computes the answer. Note that Solver might recommend a fractional part of a dollar for some answers.

Constants Section

Your spreadsheet should have the constants shown in Figure 9-6. An explanation of the line items follows the figure.

	A	B
8	**Constants**	
9	Cash available to invest	$ 100,000,000
10	Minimum investment in each type of fund	$ 10,000,000
11	Expected Rates of Return:	
12	Stock funds	8.0%
13	Bond funds	4.0%
14	Precious Metals funds	10.0%
15	REIT funds	12.0%
16	Administrative Cost Rate:	
17	Stock funds	0.8%
18	Bond funds	0.2%
19	Precious Metals funds	0.5%
20	REIT funds	0.7%
21	Risk Factor:	
22	Stock funds	3
23	Bond funds	1
24	Precious Metals funds	5
25	REIT funds	4

Source: Microsoft product screenshots used with permission from Microsoft Corporation

FIGURE 9-6   Constants section

- Cash available to invest—You have $100 million to invest.
- Minimum investment in each type of fund—You must invest at least $10 million in each type of mutual fund.
- Expected Rates of Return—The expected average rate of return is different for each kind of fund. For example, the rate is 8 percent for funds that invest in industrial company common stocks.
- Administrative Cost Rate—The expected average expense rate is different for each kind of fund. For example, the rate is eight-tenths of one percent for funds that invest in industrial company common stocks.
- Risk Factor—The expected average risk factor is different for each kind of fund. For example, the factor is 3 for funds that invest in industrial company common stocks.

Calculations Section

Your spreadsheet should calculate the amounts shown in Figure 9-7. These values will be used later in the income statement or as constraints. Calculated values may be based on the values of the changing cells, the constants, and other calculations. An explanation of the line items follows Figure 9-7.

	A	B
27	**Calculations**	
28	Total Amount Invested	
29	Percentage invested in Stocks and Bonds	
30	Revenue earned:	
31	Stock funds	
32	Bond funds	
33	Precious Metals funds	
34	REIT funds	
35	Administrative Costs:	
36	Stock funds	
37	Bond funds	
38	Precious Metals funds	
39	REIT funds	
40	Risk Points:	
41	Stock funds	
42	Bond funds	
43	Precious Metals funds	
44	REIT funds	
45	Total Risk Points to Total amount Invested	

Source: Microsoft product screenshots used with permission from Microsoft Corporation

FIGURE 9-7   Calculations section

- Total Amount Invested—Amounts invested are shown in the changing cells.

- Percentage invested in Stocks and Bonds—This value is the sum of amounts invested in stock and bond funds versus the total invested in all kinds of funds. Note that precious metals funds are not included in stock funds.
- Revenue earned—The revenue earned for a type of fund is a function of the amount invested in it and its average rate of return.
- Administrative Costs—The administrative cost for a type of fund is a function of the amount invested in it and its average administrative expense rate.
- Risk Points—The risk points for a type of fund is a function of the amount invested in it and its assigned risk factor.
- Total Risk Points to Total amount Invested—This total is the average risk factor for the fund of funds. It is the total risk points for all types of funds divided by the total amount invested in all types of funds.

Income Statement Section

The statement shown in Figure 9-8 is the projected net annual income on the amounts invested in the fund of funds. An explanation of the line items follows Figure 9-8.

	A	B
47	**Income Statement**	
48	Total Revenue	
49	Total Administrative Costs	
50	Net Income Before Tax	

Source: Microsoft product screenshots used with permission from Microsoft Corporation

FIGURE 9-8   Income Statement section

- Total Revenue—This amount is the sum of the revenues earned, as shown in the Calculations section.
- Total Administrative Costs—This amount is the sum of administrative costs, as shown in the Calculations section.
- Net Income Before Tax—This amount is the difference between Total Revenue and Total Administrative Costs.

Constraints and Running Solver

Next, you must determine the constraints to implement management's investment rules. Use Solver to enter the decision constraints for the base case. You want to maximize net income before taxes, subject to the various investment and risk constraints. Run Solver, and ask for the Answer Report when Solver reports a solution that satisfies the constraints. (For guidance on using Solver, refer to Tutorial D and Case 8.)

When you finish, print the entire spreadsheet, including the Solver Answer Report sheet. Save the file one more time using the Save command in the File tab; you can keep FundOfFundsBase.xlsx as the filename. Then, to prepare for the extension case, use the Save As command in the File tab to create a *new* spreadsheet named **FundOfFundsExtension.xlsx**.

Assignment 1B: Creating the Spreadsheet for the Extension Case

Next, you prepare the extension case. Some members of your management team think that interest rates are about to rise significantly. This increase would affect the overall investment environment, and is the main difference between the base case and the extension case.

Rising interest rates on newly issued debt would cause the market prices of existing debt to fall, which would create a lower rate of return for the bond funds owned by your fund of funds. Risk factors associated with bond funds would be expected to change.

Typically, rising interest rates also lead to lower common stock prices in the long run. Thus, you would expect a significantly lower rate of return on the stock funds owned by your fund of funds.

Costs associated with analyzing stocks and bonds would also change as analysts learn about changes in market conditions.

Figure 9-9 shows the average rates of returns that you expect from the kinds of funds in which you invested for the extension case.

Type of Fund	Expected Average Rate of Return (%)
Stocks of industrial companies	3
High-grade bonds	3
Real estate companies (REITs)	12
Precious metals companies	12

Source: © 2015 Cengage Learning®

FIGURE 9-9   Expected average rates of return for extension case

The expected administrative cost rates for each kind of fund in the extension case are shown in Figure 9-10.

Type of Fund	Expected Average Administrative Cost Rate
Stocks of industrial companies	.012 (1.2%)
High-grade bonds	.012 (1.2%)
Real estate companies (REITs)	.007 (0.7%)
Precious metals companies	.010 (1.0%)

Source: © 2015 Cengage Learning®

FIGURE 9-10   Expected average administrative cost rates for extension case

Assigned risk factors for the extension case are shown in Figure 9-11.

Type of Fund	Risk Factor
Stocks of industrial companies	3
High-grade bonds	3
Real estate companies (REITs)	5
Precious metals companies	4

Source: © 2015 Cengage Learning®

FIGURE 9-11   Assigned risk factors for extension case

Some of management's rules for the base case must be adjusted in the extension case. The average risk factor limit increases to 3.5 in recognition of increasing market uncertainties. Also, you no longer need to invest 60 percent in stock and bond funds; that rule can be dropped.

Modify the extension case spreadsheet to handle the changes. Change the constant values and constraints as needed. Run Solver, and then ask for the Answer Report when Solver reports a solution that satisfies the constraints. When you finish, print the entire spreadsheet, including the Solver Answer Report sheet. Save the file one more time, close it, and exit Excel.

ASSIGNMENT 2: USING THE SPREADSHEET FOR DECISION SUPPORT

You have built models for the base case and extension case because you want to know the investment mix for each scenario and which scenario yields the higher net income before taxes, consistent with perceived risks. You will now complete the case by (1) using the Answer Reports to gather the data you need to make the investment mix decisions and (2) documenting your recommendations in a memo.

Assignment 2A: Using the Spreadsheet to Gather Data

You have printed the Answer Report worksheets for each scenario. Each sheet reports how much of each kind of fund to purchase, the expected net income before taxes in each case, and the weighted average risk ratio in each case. Your management wants to know how much the extension case results differ from the base case results.

Assignment 2B: Documenting Your Recommendation in a Memo

Write a brief memo to the company's management that explains your results. Observe the following requirements:

- Set up your memo as described in Tutorial E.
- In the first paragraph, briefly outline the investment mix decision and the purpose of your analysis.
- Tell management which scenario yields the higher net income before taxes and which yields the lower overall risk ratio. Tell management how a rise in interest rates would be expected to change net income before taxes and the overall expected risk level. Tell management how investment allocations would differ by comparing the base case with the extension case.
- Support your statements by including a summary data table like the one shown in Figure 9-12.

	Base Case	Extension Case
Amount invested in stock funds		
Amount invested in bonds		
Amount invested in precious metals funds		
Amount invested in REITs		
Net income before taxes		
Weighted average risk ratio		

Source: © 2015 Cengage Learning®

FIGURE 9-12   Format of table to enter in memo

ASSIGNMENT 3: GIVING AN ORAL PRESENTATION

Assume that your boss is very impressed with your work and has asked you to give an oral presentation of your findings to the company's senior management. Assume that management wants the presentation to last no more than 10 minutes. Use visual aids or handouts that you think are appropriate. See Tutorial F for tips on preparing and giving an oral presentation.

DELIVERABLES

Your completed case should include the following deliverables for your instructor:

- A printed copy of your memo
- Printouts of your spreadsheets and Solver answer sheets
- Presentation visual aids and handouts as appropriate
- Electronic copies of your memo and Excel files

PART 4

DECISION SUPPORT CASE
USING BASIC EXCEL FUNCTIONALITY

CASE 10
Golden Retriever Recovery Agency, 207

GOLDEN RETRIEVER RECOVERY AGENCY

Decision Support Using Microsoft Excel

PREVIEW

A regional bank outsources its asset recovery operations for past-due automobile loans to an agency that specializes in the car industry. Golden Retriever Recovery Agency uses various methods to recover assets, including vehicle repossessions, collection efforts by telephone, and letters reminding customers to pay. Generally speaking, recovery efforts have three possible outcomes: Either the customer pays, the car is repossessed, or the customer settles the account for a lesser amount. In this case, you will use Excel to summarize the campaign results of a collections letter sent to customers whose accounts are past due and determine whether the letter campaign was successful in increasing the amount of recovered assets.

PREPARATION

- Review spreadsheet concepts discussed in class and in your textbook.
- Complete any exercises that your instructor assigns.
- Review file-saving procedures for Windows programs.
- Refer to Tutorials E and F as necessary.

BACKGROUND

The mission of Golden Retriever Recovery Agency is to help car loan originators mitigate risk and reduce losses on bad loans. A bad loan is one for which customers have stopped making payments because of unemployment, divorce, medical problems, reduction of income, or death in the family. As a recovery analyst for the agency, your job is to track the performance of Golden Retriever's settlement campaigns to determine if they are effective in mitigating risk and reducing bad loan balances. In a settlement campaign, customers are mailed a letter with an offer to pay a lesser amount than they currently owe and a request to call the customer service phone line to complete the process.

When a customer is late making payments and there is a high risk that the customer might stop paying altogether, Golden Retriever can make a settlement offer for less than the amount owed. Although a settlement would not recover the entire loan balance, this option is often better than repossessing the vehicle if the settlement amount is greater than the vehicle's market value. For example, if a customer is late making payments on a car loan that has a $10,000 balance and the current market price of the car is $5,000, the bank may offer a settlement for 75 percent, or $7,500. Repossession of the vehicle would result in a $5,000 loss, whereas a 75% settlement would reduce the loss to $2,500. Although both options carry negative implications for the customer, a settlement is seen more favorably by consumer credit agencies that monitor credit scores for customers. On the other hand, if a past-due customer owes $10,000 and the car's current market price is $12,000, the bank may choose to repossess the car. The repossession would allow the bank to sell the car, eliminate the loan balance, and possibly turn a profit on the sale.

Locating and contacting customers whose accounts are past due can be difficult. Contact rates deteriorate as the customer's count of past due days increases. It is much easier to contact a customer whose account is 10 days past due than one whose account is 120 days past due.

You are tracking the performance of a recovery campaign that has mailed notices to thousands of customers who are 90 to 150 days past due and are considered high risks. The campaigns are a good opportunity to test different collection tactics; for this purpose you have included two templates.

Template A approaches the customer from an educational point of view and explains the negative consequences of not paying off loans. Template B takes a more aggressive tone, reminding customers of the legal implications of not paying off the loan.

The campaign's budget is $10,000, and the cost of one mail piece is $1.41. The goals of the campaign are listed below:

- Inform customers in the letter that settlement options are available.
- Generate inbound calls to the customer service line to increase contact rates.
- Test two templates (A and B) to see which one performs better.

In addition to head-to-head testing of the templates, the campaign includes "hold out" groups for control purposes. The purpose of the control groups is to test the overall effectiveness of the campaign by comparing them with the performance of the templates. Accounts in the control group are *not* mailed any letters; these accounts make up about 20 percent of the total group. You also want to track performance by the accounts' days past due, so you split the groups (Template A, Template B, and Control) into 90 to 120 days past due and 121 to 150 days past due. To track the campaign's performance, you set up the campaign according to Figure 10-1.

Test Group	Proportion	Quantity
Template A 90–120 Days Past Due	~20%	1728
Template A 121–150 Days Past Due	~20%	1742
Template B 90–120 Days Past Due	~20%	1777
Template B 121–150 Days Past Due	~20%	1716
Control 90–120 Days Past Due	~10%	792
Control 121–150 Days Past Due	~10%	833

Source: © 2015 Cengage Learning®

FIGURE 10-1   Campaign test setup

The campaign's performance will be measured using four different rates:

- Unit Pay Rate—This rate measures the campaign's performance as a function of the number of account holders who make a payment versus the total number of accounts in the campaign.
- Dollar Pay Rate—This rate measures the campaign's performance as a function of the sum of all payments made versus the sum of all balances for all accounts in the campaign.
- Settlement Rate—This rate measures the campaign's performance as a function of the number of account holders who accept a settlement versus the total number of accounts in the campaign.
- Inbound Rate—This rate measures the campaign's performance as a function of the number of account holders who call the customer service phone line versus the total number of accounts in the campaign.

ASSIGNMENT 1: CREATING A SPREADSHEET FOR DECISION SUPPORT

In this assignment, you produce a spreadsheet that will help you track the campaign's performance. In Assignment 2, you will summarize the campaign's results using pivot tables. In Assignment 3, you may be asked to prepare an oral presentation of your analysis.

A spreadsheet skeleton has been started and is available for you to use; it will save you time. If you want to use the spreadsheet skeleton, locate Case 10 in your data files and then select

GoldenRetrieverRecoveryAgency.xlsx. Note that the main tab of the spreadsheet is named Skeleton. Your spreadsheet should contain the following sections:

- Constants
- Inputs
- Summary of Key Results
- Calculations

A discussion of each section follows.

Constants Section

Your spreadsheet will include the constants shown in Figure 10-2. An explanation of each line item follows the figure.

	A	B
1	**GOLDEN RETRIEVER RECOVERY AGENCY**	
2		
3	**CONSTANTS**	
4	Cost per mail piece	$ 1.41
5	Template A 90-120 Volume	1728
6	Template B 90-120 Volume	1777
7	Template A 121-150 Volume	1742
8	Template B 121-150 Volume	1716
9	Control 90-120 Volume	792
10	Control 121-150 Volume	833

Source: Microsoft product screenshots used with permission from Microsoft Corporation

FIGURE 10-2   Constants section

- Cost per mail piece—This value is the total cost of mailing one letter.
- Template A 90–120 Volume—This value is the total number of accounts included in the template A test group that are 90 to 120 days past due.
- Template B 90–120 Volume—This value is the total number of accounts included in the template B test group that are 90 to 120 days past due.
- Template A 121–150 Volume—This value is the total number of accounts included in the template A test group that are 121 to 150 days past due.
- Template B 121–150 Volume—This value is the total number of accounts included in the template B test group that are 121 to 150 days past due.
- Control 90–120 Volume—This value is the total number of accounts included in the control group that are 90 to 120 days past due.
- Control 121–150 Volume—This value is the total number of accounts included in the control group that are 121 to 150 days past due.

Inputs Section

Your spreadsheet should include the inputs you calculate using pivot tables in Assignment 2. The inputs are shown in Figure 10-3. An explanation of the line items follows the figure.

13	INPUTS						
14		Group	Balance	Payments	Pay Collected	Completed Settlement	Inbound
15	90-120 Days Past Due	Template A					
16		Template B					
17		Total Test					
18		Control					
19							
20		Group	Balance	Payments	Pay Collected	Completed Settlement	Inbound
21	121-150 Days Past Due	Template A					
22		Template B					
23		Total Test					
24		Control					
25							
26		Group	Balance	Payments	Pay Collected	Completed Settlement	Inbound
27	Overall	Template A					
28		Template B					
29		Total Test					
30		Control					

Source: Microsoft product screenshots used with permission from Microsoft Corporation

FIGURE 10-3   Inputs section

- Balance—This column shows the total outstanding balances of overdue accounts. Format the cells as currency with no decimals.
  - 90–120 Days Past Due—This value is calculated using pivot tables in Assignment 2.
  - 121–150 Days Past Due—This value is calculated using pivot tables in Assignment 2.
  - Overall—This value is calculated using pivot tables in Assignment 2.
- Payments—This column shows the number of payments collected. Format the cells as numbers with no decimals.
  - 90–120 Days Past Due—This value is calculated using pivot tables in Assignment 2.
  - 121–150 Days Past Due—This value is calculated using pivot tables in Assignment 2.
  - Overall—This value is calculated using pivot tables in Assignment 2.
- Pay Collected—This column shows the total number of dollars collected on the accounts. Format the cells as currency with no decimals.
  - 90–120 Days Past Due—This value is calculated using pivot tables in Assignment 2.
  - 121–150 Days Past Due—This value is calculated using pivot tables in Assignment 2.
  - Overall—This value is calculated using pivot tables in Assignment 2.
- Completed Settlement—This column shows the number of customers who requested and completed a settlement. Format the cells as numbers with no decimals.
  - 90–120 Days Past Due—This value is calculated using pivot tables in Assignment 2.
  - 121–150 Days Past Due—This value is calculated using pivot tables in Assignment 2.
  - Overall—This value is calculated using pivot tables in Assignment 2.
- Inbound—This column shows the total number of inbound calls for each group. Customers made these calls to the customer service line. Format the cells as numbers with no decimals.
  - 90–120 Days Past Due—This value is calculated using pivot tables in Assignment 2.
  - 121–150 Days Past Due—This value is calculated using pivot tables in Assignment 2.
  - Overall—This value is calculated using pivot tables in Assignment 2.

Summary of Key Results Section

Your spreadsheet should include the key results shown in Figure 10-4. The values are echoed from other parts of your spreadsheet. An explanation of the line items follows the figure.

32 SUMMARY OF KEY RESULTS	Unit Pay Rate	Dollar Pay Rate	Settlement Rate	Inbound Rate
33 Overall Template A				
34 Overall Template B				
35 Overall Control				

Source: Microsoft product screenshots used with permission from Microsoft Corporation

FIGURE 10-4   Summary of Key Results section

[handwritten annotation: A = # payments / # Accounts]

[handwritten annotation left margin: Total Payments / Total Bal]

- Unit Pay Rate—This value is the number of all payments in a group divided by the total number of accounts in the same group. The three groups are Template A, Template B, and the control group.
- Dollar Pay Rate—This value is the dollar sum of all payments in a group divided by the sum of all balances in the same group.
- Settlement Rate—This value is the number of all settlement payments in a group divided by the total number of accounts in the same group. *[handwritten: Set / # act.]*
- Inbound Rate—This value is the number of calls to date in a group divided by the total number of accounts in the same group. *[handwritten: # calls / # accts.]*

Calculations Section

Your worksheet should include the calculations included in Figure 10-5. An explanation of the line items follows the figure.

37 CALCULATIONS								
38	Group	Volume	Campaign Cost	Unit Pay Rate	Dollar Pay Rate	Settlement Rate	Inbound Rate	
39 90-120 Days Past Due	Template A							
40	Template B							
41	Total Test							
42	Control							
43								
44	Group	Volume	Campaign Cost	Unit Pay Rate	Dollar Pay Rate	Settlement Rate	Inbound Rate	
45 121-150 Days Past Due	Template A							
46	Template B							
47	Total Test							
48	Control							
49								
50	Group	Volume	Campaign Cost	Unit Pay Rate	Dollar Pay Rate	Settlement Rate	Inbound Rate	
51 Overall	Template A							
52	Template B							
53	Total Test							
54	Control							

Source: Microsoft product screenshots used with permission from Microsoft Corporation

FIGURE 10-5   Calculations section

[handwritten annotation: use constants as references? do values need to be typed in (getpivot)]

- Volume—This column shows the number of accounts included in each group. Format the cells as numbers with no decimals.
  - 90–120 Days Past Due
    - Template A—This value is echoed from the Constants section.
    - Template B—This value is echoed from the Constants section.
    - Total Test—This value is the sum of the preceding values for templates A and B.
    - Control—This value is echoed from the Constants section.
  - 121–150 Days Past Due
    - Template A—This value is echoed from the Constants section.
    - Template B—This value is echoed from the Constants section.
    - Total Test—This value is the sum of the preceding values for templates A and B.
    - Control—This value is echoed from the Constants section.
  - Overall
    - Template A—This value is the sum of volume from accounts that are 90–120 days past due and 121–150 days past due for template A.

- Template B—This value is the sum of volume from accounts that are 90–120 days past due and 121–150 days past due for template B.
- Total Test—This value is the sum of the preceding values for templates A and B.
- Control—This value is the sum of volume from control accounts that are 90–120 days past due and 121–150 days past due.
- Campaign Cost—This column shows the total cost of mailing the letters. Format the cells as currency with two decimals.
  - 90–120 Days Past Due
    - Template A—This value is calculated by multiplying the cost per mail piece in the Constants section by the volume of accounts in the group.
    - Template B—This value is calculated by multiplying the cost per mail piece in the Constants section by the volume of accounts in the group.
    - Total Test—This value is the sum of the preceding values for templates A and B.
    - Control—No calculation is needed.
  - 121–150 Days Past Due
    - Template A—This value is calculated by multiplying the cost per mail piece in the Constants section by the volume of accounts in the group.
    - Template B—This value is calculated by multiplying the cost per mail piece in the Constants section by the volume of accounts in the group.
    - Total Test—This value is the sum of the preceding values for templates A and B.
    - Control—No calculation is needed.
  - Overall
    - Template A—This value is the sum of costs from accounts that are 90–120 days past due and 121–150 days past due for template A.
    - Template B—This value is the sum of costs from accounts that are 90–120 days past due and 121–150 days past due for template B.
    - Total Test—This value is the sum of the preceding values for templates A and B.
    - Control—No calculation is needed.
- Unit Pay Rate—This column shows the ratio of payments to accounts in the same group. Format the cells as percentages with two decimals.
  - 90–120 Days Past Due—This value is calculated by dividing the number of payments by the accounts in the group.
  - 121–150 Days Past Due—This value is calculated by dividing the number of payments by the accounts in the group.
  - Overall—This value is calculated by dividing the number of payments by the accounts in the group.
- Dollar Pay Rate—This column shows the ratio of the sum of total payments to the sum of all balances per group. Format the cells as percentages with two decimals.
  - 90–120 Days Past Due—This value is the sum of payment amounts per group divided by the balances in the same group.
  - 121–150 Days Past Due—This value is the sum of payment amounts per group divided by the balances in the same group.
  - Overall—This value is the sum of payment amounts per group divided by the balances in the same group.
- Settlement Rate—This column shows the ratio of the number of settlements to total accounts in each group. Format the cells as percentages with two decimals.
  - 90–120 Days Past Due—This value is the number of completed settlements per group divided by the volume of accounts in the same group.
  - 121–150 Days Past Due—This value is the number of completed settlements per group divided by the volume of accounts in the same group.
  - Overall—This value is the number of completed settlements per group divided by the volume of accounts in the same group.
- Inbound Rate—This column shows the ratio of the number of inbound calls to the total accounts in the group. Format the cells as percentages with two decimals.
  - 90–120 Days Past Due—This value is the number of inbound calls per group divided by the volume of accounts in the same group.

- 121–150 Days Past Due—This value is the number of inbound calls per group divided by the volume of accounts in the same group.
- Overall—This value is the number of inbound calls per group divided by the volume of accounts in the same group.

ASSIGNMENT 2: USING THE SPREADSHEET FOR DECISION SUPPORT

You will now complete the case by (1) using the spreadsheet to summarize the campaign data using pivot tables, (2) documenting your findings in a memo, and (3) giving an oral presentation if your instructor requires it.

Assignment 2A: Using Pivot Tables to Gather Balance Information

The List of Accounts tab contains all of the accounts included in the campaign. The tab also includes details such as outstanding balances, days past due, and the template used. You will use a pivot table to summarize the total balance amounts for each of the required groups in the Inputs section. First, you must create a new column to calculate whether the accounts' days past due fall into the range of 90–120 days or 121–150 days. The simplest way to make this calculation is to use an If statement. Use Figure 10-6 to help you get started. When you have calculated the group data for each account, you can create the pivot table to summarize the balance data.

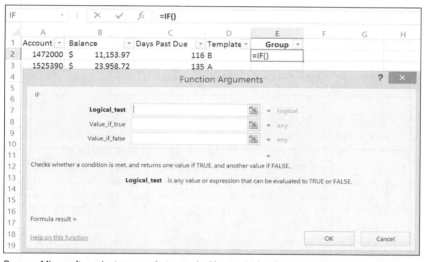

Source: Microsoft product screenshots used with permission from Microsoft Corporation

FIGURE 10-6   Creating an If statement

Insert a pivot table into a new worksheet named **Balances**. Using a new worksheet provides a cleaner environment for working with the data and avoids the possibility that the original data will become corrupted. Figure 10-7 shows the first step of creating the pivot table. You can also refer to Tutorial E for more tips on creating pivot tables.

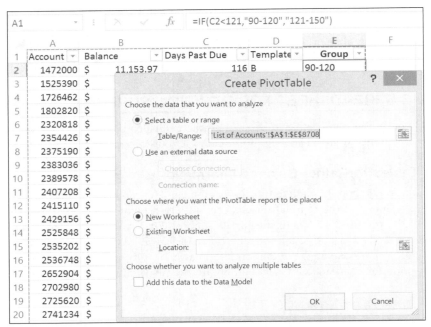

Source: Microsoft product screenshots used with permission from Microsoft Corporation

FIGURE 10-7   Inserting a pivot table into the new worksheet

Using the pivot table, calculate the sum of all balances per group. Insert the results into the appropriate Balance cells in the Inputs section of the Skeleton tab.

Assignment 2B: Using Pivot Tables to Gather Payment Information

The List of Payments tab contains all of the accounts included in the campaign. The tab also includes details such as transactions, amounts, days past due, and the template used. You will use a pivot table to summarize the total payment amounts and total number of payments for each of the required groups in the Inputs section. First, you must create a new column to calculate whether the accounts' days past due fall into the range of 90–120 days or 121–150 days. The simplest way to make this calculation is to use an If statement. Use Figure 10-6 to help you get started. When you have calculated the group data for each account, you can create the pivot table to summarize the payment data.

Insert a pivot table into a new worksheet named **Payments**. As mentioned, using a new worksheet provides a cleaner environment for working with the data and avoids the possibility that the original data will become corrupted. If necessary, refer back to Figure 10-7 to see the first step of creating the pivot table.

To count the number of payments, use the Transaction field in the List of Payments tab. When a payment is received, it is noted with a transaction code of 300. To sum the payment amounts, use the Amount field. Insert the number of payments into the Payments cells and the payment amounts into the Pay Collected cells in the Inputs section of the Skeleton tab.

Assignment 2C: Using Pivot Tables to Gather Settlement Information

The List of Settlements tab contains all of the accounts included in the campaign. The tab also includes details such as account status, days past due, and the template used. You will use a pivot table to summarize the total number of settlements for each of the required groups in the Inputs section. First, you must create a new column to calculate whether the accounts' days past due fall into the range of 90–120 days or 121–150 days. Make this calculation for each account using an If statement, as shown in Figure 10-6, and then create the pivot table to summarize the settlement data.

Insert the pivot table into a new worksheet named **Settlements**. See Figure 10-7 if you need help creating the pivot table.

To count the number of settlements, use the Status field in the List of Settlements tab. When a settlement is accepted, its status changes to "Completed Settlement." Insert the number of settlements into the Completed Settlement cells in the Inputs section of the Skeleton tab.

Assignment 2D: Using Pivot Tables to Gather Inbound Call Information

The List of Inbounds tab contains all of the accounts included in the campaign. The tab also includes details such as inbound call dates, days past due, and the template used. You will use a pivot table to summarize the total number of inbound calls for each of the required groups in the Inputs section. First, you must create a new column to calculate whether the accounts' days past due fall into the range of 90–120 days or 121–150 days. Make this calculation for each account using an If statement, as shown in Figure 10-6, and then create the pivot table to summarize the inbound calls data.

Insert the pivot table into a new worksheet named **Inbounds**. See Figure 10-7 if you need help creating the pivot table.

To count the number of inbound calls, use the Inbound Date field in the List of Inbounds tab. When a call is received, the date is logged. Insert the number of inbound calls into the Inbound cells in the Inputs section of the Skeleton tab.

ASSIGNMENT 3: DOCUMENTING YOUR FINDINGS

In this assignment, you write a memo in Microsoft Word that documents your findings and recommendations. Use supporting charts and figures to justify your analysis and recommendations. Based on your analysis, answer the following questions:

Executive Summary

- Did the campaign finish over budget or under budget?
- Did the test groups beat the control group in all rates, or only in some? → *Some*
- Did the campaign achieve its goal of increasing inbound rates?
- Did one template perform better than the other?
- Did the number of days past due have any impact on the rates?
- If you found differences in the groups, you may consider performing simple statistical tests to determine whether the differences are significant.
- Based on the results, would you recommend more letter campaigns? Why?

In addition to your memo, summarize your findings and recommendations in a PowerPoint document and then prepare to give an oral presentation. Tutorial F explains how to prepare and give an effective oral presentation.

DELIVERABLES

Prepare the following deliverables for your instructor:

- Completed spreadsheet
- A memo with your findings and recommendations
- A PowerPoint document for your oral presentation

do pivot tables need to be put with the skeleton?

do use volume from constants table or cal. table.

PART 5

INTEGRATION CASES USING ACCESS AND EXCEL

THE DRAFT PICK CHART ANALYSIS

Decision Support with Access and Excel

PREVIEW

A professional football league selects its new players each year from college teams in a process known as the draft. Earlier picks are more valued than later picks. Executives of the league use a chart developed by a former coach to indicate the relative value of each pick. In this case, you will use Microsoft Access and Excel to determine how well the chart's values correlate to actual performance by players on the field.

PREPARATION

- Review database and spreadsheet concepts discussed in class and in your textbook.
- Complete any exercises that your instructor assigns.
- Complete any parts of Tutorials B, C, and D that your instructor assigns, or refer to them as necessary.
- Review file-saving procedures for Microsoft Windows programs, as discussed in Tutorial C.
- Refer to Tutorials E and F as necessary.

BACKGROUND

The Continental Football League (CFL) has 20 teams. Teams get new players each year from college football teams using a process known as the draft. After the season is over, representatives from each team meet in New York City. In predetermined order, the teams draft players they want for the next season. There are five rounds of picks, which means that 100 players are drafted.

The teams intensively analyze players' college performance prior to the draft. Of course, the most highly rated players are drafted earliest.

Sometimes teams trade draft picks before the draft begins. For example, assume that the Buffalo team will have the third pick in the draft and that Jacksonville team officials covet that pick, which would enable them to draft a player they like. As it stands, Jacksonville team officials are sure the player will have already been drafted when their turn comes around. Jacksonville holds the 18th and 38th picks in the draft, and offers both picks to Buffalo in return for the third pick in the draft.

If you were drafting for Buffalo, would you make the trade? On one hand, you would gain an extra pick in the draft, which is surely an advantage. On the other hand, you'd be reluctant to relinquish that early pick—players picked very early in the draft often go on to become the league's superstars, and players picked later generally do not. How do you evaluate the Jacksonville offer?

Before the 1990 draft, legendary Los Angeles coach Alfred "Nails" Smith created a chart designed to help team executives on draft day. He assigned point values to each pick. For example, he assigned the first pick a value of 3,000 points, the 18th pick a value of 900 points, the 38th pick a value of 520 points, and so on until the 100th pick, which he gave a value of 100 points. Using this chart, Buffalo team officials can easily see that the Jacksonville offer is inadequate. The third pick is worth 2,200 points, and the Jacksonville picks together are worth only 1,420 points. Buffalo would decline the offer if team officials follow the chart's logic.

Smith's chart has been used by CFL teams all these years. Smith always refused to reveal the basis for his draft pick rankings. Now that he has died, team officials will never know what he was thinking when he created the chart.

You live in a large city that has a team in the league, and you work for the team as an analyst. The team's general manager (GM), who is in charge of the team's draft, has asked you to analyze the chart's values. The chart is shown in Figure 11-1.

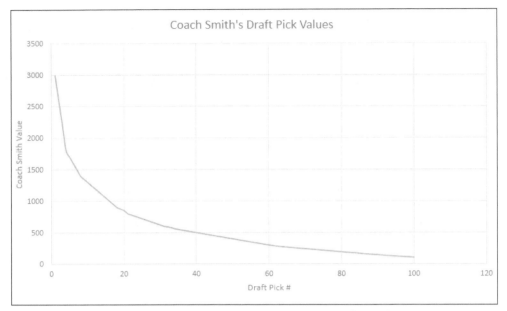

Source: Microsoft product screenshots used with permission from Microsoft Corporation

FIGURE 11-1   Draft pick value chart

"Teams use this chart all the time to assess trade offers before the draft. I want to know if a graph of actual player performance matches up with this curve," the GM says. "Of course, early picks should be more valued than later picks. But as you can see, early picks are *much* more highly valued than later picks. Is it really the case that later picks are *that* much less valuable than early picks? What if, in fact, later picks are more valuable than we think? I could trade an early pick for a lot of later picks, and actually get more value than the other team thinks it is giving up!"

The GM wants a curve based on actual performance. "Go back to the draft three years ago. Take a look at how well those 100 drafted players actually did in the league in the three years since." Pay attention to three things:

1. Was the player on a team's roster during the year? In other words, did he play?
2. Was the player a starting player for the team during the year?
3. Did the player make the league all-star team during the year?

"The most basic form of a player's value is merely making the team," she says. "Players that start are more valuable than reserve players. Of those players that start, those who make the league all-star team are the most valuable.

"Gather the data and put it into a database," she says. "Then, build me a curve that shows actual value. You need weightings to combine the three forms of value into one measurement. Let's say that being a starter gives twice the value of merely making the team. Let's say that making the all-star team gives three times more value than merely making our team. Then compare the two curves and tell me how well the coach's famous curve corresponds with real value."

Gather the data and enter it in a database called DraftChoices.accdb. The database file is available to you in the files that come with your casebook. Locate Case 11 in your data files and then select **DraftChoices.accdb**.

Using the Database

The tables in the database file are discussed next. Figure 11-2 shows the design of the Players Drafted table.

Players Drafted	
Field Name	Data Type
Draft Num	Number
Round Num	Number
Player Name	Short Text

Source: Microsoft product screenshots used with permission from Microsoft Corporation

FIGURE 11-2   Design of the Players Drafted table

Each draft pick is assigned a unique number between 1 and 100 in the Draft Num field. Draft Num is the primary key field. There are five rounds in the draft, which means each of the 20 teams makes five draft picks. The last name of each drafted player is recorded in the Player Name field.

Figure 11-3 shows a few of the records in the Players Drafted table.

Players Drafted		
Draft Num	Round Num	Player Name
1	1	Stewart
2	1	Sanchez
3	1	Morris
4	1	Rogers
5	1	Reed

Source: Microsoft product screenshots used with permission from Microsoft Corporation

FIGURE 11-3   Some data records in the Players Drafted table

For example, the first player taken in round 1 is named Stewart. The fifth player drafted in the first round is named Reed.

Figure 11-4 shows the design of the In The League table.

In The League	
Field Name	Data Type
Draft Num	Number
Year1	Number
Year2	Number
Year3	Number

Source: Microsoft product screenshots used with permission from Microsoft Corporation

FIGURE 11-4   Design of the In The League table

Draft Num is a unique number between 1 and 100, and is the primary key field. If a player was in the league in the first year after the draft, the Year1 field will have a value of 1. If a player was not in the league in the first year after the draft, the field will have a value of 0. Values in the Year2 and Year3 fields use the same logic as values in the Year1 field.

Figure 11-5 shows a few records in the In The League table.

In The League			
Draft Num	Year1	Year2	Year3
1	1	1	1
2	1	1	1
3	1	1	1
4	1	1	1
5	1	1	1

Source: Microsoft product screenshots used with permission from Microsoft Corporation

FIGURE 11-5   Some data records from the In The League table

For example, the player taken with the first draft pick was in the league in each of the three years. If the player had not been in the league during the third year after the draft, the Year3 field would have a value of 0.

Figure 11-6 shows the design of the Starter In Year table.

Starter In Year	
Field Name	Data Type
Draft Num	Number
Year1	Number
Year2	Number
Year3	Number

Source: Microsoft product screenshots used with permission from Microsoft Corporation

FIGURE 11-6   Design of the Starter In Year table

Draft Num is a unique number between 1 and 100, and is the primary key field. If a player was in his team's starting lineup in the first year after the draft, the Year1 field will have a value of 1. If a player was not a starter in the first year after the draft, the field will have a value of 0. Values in the Year2 and Year3 fields use the same logic as values in the Year1 field.

Figure 11-7 shows a few records in the Starter In Year table.

Starter In Year			
Draft Num ▾	Year1 ▾	Year2 ▾	Year3 ▾
1	1	1	1
2	1	1	1
3	1	1	1
4	1	1	1
5	1	1	1
6	1	1	0

Source: Microsoft product screenshots used with permission from Microsoft Corporation

FIGURE 11-7   Some data records in the Starter In Year table

For example, the player taken with the sixth pick in the draft three years ago was a starter in his first two years, but was not a starter for his team in the third year after the draft.

Figure 11-8 shows the design of the All Star In Year table.

All Star In Year	
Field Name	Data Type
Draft Num	Number
Year1	Number
Year2	Number
Year3	Number

Source: Microsoft product screenshots used with permission from Microsoft Corporation

FIGURE 11-8   Design of the All Star In Year table

Draft Num is a unique number between 1 and 100, and is the primary key field. If a player was a league all-star in the first year after the draft, the Year1 field will have a value of 1. If a player was not an all-star in the first year after the draft, the field will have a value of 0. Values in the Year2 and Year3 fields use the same logic as values in the Year1 field.

Figure 11-9 shows a few records in the All Star In Year table.

All Star In Year			
Draft Num ▾	Year1 ▾	Year2 ▾	Year3 ▾
1	1	1	1
2	1	1	0
3	0	0	0
4	0	0	0
5	0	0	0
6	1	1	0

Source: Microsoft product screenshots used with permission from Microsoft Corporation

FIGURE 11-9   Some data records in the All Star In Year table

For example, the player taken with the first draft pick was an all-star in each of the three seasons after the draft three years ago. The player taken with the third pick was never an all-star. The player taken with the sixth pick was an all-star in his first two years, but was not in the third year after the draft.

Figure 11-10 shows the design of the Coaches Values table.

Coaches Values	
Field Name	Data Type
Pick Num	Number
Value	Number

Source: Microsoft product screenshots used with permission from Microsoft Corporation

FIGURE 11-10   Design of the Coaches Values table

Pick Num is a unique number between 1 and 100, and is the primary key field. The Value field holds the value assigned by Coach Smith when he created the chart.

Figure 11-11 shows a few records in the Coaches Values table.

Coaches Values

Pick Num	Value
1	3000
2	2600
3	2200
4	1800
5	1700
6	1600

Source: Microsoft product screenshots used with permission from Microsoft Corporation

FIGURE 11-11 Some data records in the Coaches Values table

For example, the coach assigned 3,000 points to the draft's first pick and 1,600 points to the sixth pick.

ASSIGNMENT 1: USING ACCESS AND EXCEL FOR DECISION SUPPORT

The GM wants to compare actual player value with Coach Smith's pre-draft values. She has some queries in mind for measuring the value of draft picks. After you create the queries in Access, you will import some Access data into Excel for further analysis.

Queries for Measuring Draft Pick Values

The GM wants to verify that picks in early rounds are more valuable than picks in later rounds. She wants you to create three queries.

Query 1

The GM wants a query that computes the total number of years draft picks played in the league, round by round. Presumably, players selected in round 1 were on teams longer than players selected in round 2, players selected in round 2 were on teams longer than players selected in round 3, and so on. Create a query that displays the data. The output should look like the data in Figure 11-12.

Total In League

Round Num	Total Years
1	56
2	52
3	51
4	46
5	35

Source: Microsoft product screenshots used with permission from Microsoft Corporation

FIGURE 11-12 Data for total years in league, by round

Notice that the 20 people drafted in the first round almost always made a team's roster in the three years. (If all players had made a team in each of the years, Total Years would be 60.) The 20 people drafted in the second round were almost as successful as those drafted in the first round. Save the query as Total In League.

Query 2

In a similar vein, the GM wants a query that computes the total number of years draft picks were starters, round by round. Presumably, more players selected in round 1 were starters than players selected in round 2, more players in round 2 were starters than players selected in round 3, and so on. Create a query that displays the data. The output should look like the data in Figure 11-13.

Total Starters

Round Num	Total Starters
1	52
2	41
3	37
4	27
5	13

Source: Microsoft product screenshots used with permission from Microsoft Corporation

FIGURE 11-13 Data for total years as a starter, by round

Save the query as Total Starters.

Query 3

Sometimes a player drafted in a late round turns out to be very good. The GM wants data for how often this happens. She tells you to develop a simple player rating using an unweighted approach. The most points possible for this query would be 9: A player might make a team for three years, start for a team for three years, and be an all-star for three years. Create a query to rank players by total point value. Your output should look like that shown in Figure 11-14.

Draft Num	Round Num	Player Name	Points
12	1	Cooper	9
10	1	Bailey	9
1	1	Stewart	9
15	1	Howard	8
2	1	Sanchez	8
8	1	Bell	7
21	2	James	7
25	2	Sanders	7
32	2	Coleman	7

Source: Microsoft product screenshots used with permission from Microsoft Corporation

FIGURE 11-14   Total points for each player drafted

Note that only three players achieved a perfect score of 9, and that they were the first, tenth, and twelfth players selected in the draft. The output is sorted by points. The GM can easily see how often late draft choices do as well as early choices by quickly scanning the output. Save the query as Total Points.

Think about how to design the Total Points query. One approach is to base it on three other queries, each of which determines a total. Then, Total Points could merely sum the values from the other three queries. For example, one of the queries could compute total starter points for each player drafted. The output of such a query would look like that shown in Figure 11-15.

Draft Num	Starter
1	3
2	3
3	3
4	3
5	3
6	2
7	2
8	3

Source: Microsoft product screenshots used with permission from Microsoft Corporation

FIGURE 11-15   Points accumulated as a starting player

For example, the player drafted with the first pick started for a team in all three years. The player taken with the sixth pick started for a team in two of the three years.

When you finish the queries, close the database file and exit Access.

Importing Data into Excel for Further Analysis

You will build a player performance value curve in Excel. To prepare, you will import four tables into Excel worksheets. First, open a new file in Excel and save it as **DraftChoices.xlsx**.

Import the Coaches Values table into Excel. To begin, select cell A1. Then, click the Data tab and select From Access in the Get External Data group. Specify the Access filename, and then specify the table to import (Coaches Values). Next, specify where to place the data in the worksheet (cell A1). The data is imported into Excel as a data table, which is not the format you want. In the Tools group, select Convert To Range. Rename the worksheet as Coaches Values.

In the same way, import the data from the All Star In Year, Starter In Year, and In The League tables into Excel. Dedicate one worksheet to each table, converting to a range in each case. Rename worksheets after the table names.

Creating Data Needed for Charting

Your goal is to create an Excel chart that shows Coach Smith's old value curve versus an actual curve of player performance. Notice that the coach's data values and the player performance data values have different arbitrary bases, so they are not directly comparable. In each case, the data will be expressed as a percentage of the total; the percentages then can be compared directly.

Smith's pick values must be expressed as a percentage of all pick values. Therefore, divide each pick's value by the sum of all 100 pick values. When you finish, the first few rows in the Coaches Values worksheet should look like Figure 11-16.

	A	B	C
1	Pick Num	Value	Relative Value
2	1	3000	5.28%
3	2	2600	4.58%
4	3	2200	3.87%
5	4	1800	3.17%
6	5	1700	2.99%

Source: Microsoft product screenshots used with permission from Microsoft Corporation

FIGURE 11-16   Rows in the Coaches Values worksheet

Notice that each pick's value is shown relative to the sum of all values. As a check, use the SUM() function to verify that all relative values add to 100.00%.

In The League data, starter data, and all-star data must be properly weighted and then combined. (Recall that In The League data tracks whether players were in the league in the years after the draft.) To begin, open the In The League worksheet and multiply each pick's total points by 10. When you finish, the first few rows of the In The League worksheet should look like Figure 11-17.

	A	B	C	D	E
1	Draft Num	Year1	Year2	Year3	In League Points
2	1	1	1	1	30
3	2	1	1	1	30
4	3	1	1	1	30
5	4	1	1	1	30

Source: Microsoft product screenshots used with permission from Microsoft Corporation

FIGURE 11-17   Rows of the In The League worksheet

Being a starter is twice as valuable as merely being an active player in the league. In the Starter In Year worksheet, each draft pick's total points should be multiplied by 20. When you finish, the first few rows in the Starter In Year worksheet should look like Figure 11-18.

	A	B	C	D	E
1	Draft Num	Year1	Year2	Year3	Starter Points
2	1	1	1	1	60
3	2	1	1	1	60
4	3	1	1	1	60
5	4	1	1	1	60

Source: Microsoft product screenshots used with permission from Microsoft Corporation

FIGURE 11-18   Rows in the Starter In Year worksheet

Being an all-star is three times more valuable than being a regular player in the league. In the All Star In Year worksheet, multiply each draft pick's total points by 30. When you finish, the first few rows in the All Star In Year worksheet should look like Figure 11-19.

	A	B	C	D	E
1	Draft Num	Year1	Year2	Year3	All Star Points
2	1	1	1	1	90
3	2	1	1	0	60
4	3	0	0	0	0
5	4	0	0	0	0

Source: Microsoft product screenshots used with permission from Microsoft Corporation

FIGURE 11-19   Rows in the All Star In Year worksheet

In The League data, starter data, and all-star data must now be combined. Open a new worksheet and rename it as Summary. Copy all columns that contain totals to the Summary worksheet. (To paste a copy of a value derived by formula, click the Home tab, click the down arrow on the Paste button, click Paste Special in the menu, and then click the Values button in the Paste Special window. Click OK.) In the Summary worksheet, the values should be totaled and then expressed as relative values. When you finish, the first few rows in the Summary worksheet should look like Figure 11-20.

	A	B	C	D	E	F
1	Draft Num	In The League	Starter In Year	All Star In Year	Total Points	Relative
2	1	30	60	90	180	2.77%
3	2	30	60	60	150	2.31%
4	3	30	60	0	90	1.39%
5	4	30	60	0	90	1.39%
6	5	30	60	0	90	1.39%
7	6	20	40	60	120	1.85%

Source: Microsoft product screenshots used with permission from Microsoft Corporation

FIGURE 11-20   Rows in the Summary worksheet

A data series may exhibit variability, making it difficult to see the underlying trend. A moving average of the data can be used to eliminate some variability by "smoothing out" the data series. As an example, assume you have data for 200 days of a common stock index's closing values. The index values fluctuate, of course, so you want to eliminate some of the noise and see the underlying trend more clearly. You could compute a five-day moving average—for example, the value for the fifth day would be the sum of the index values for the first five days, divided by five. The value for the sixth day would be the summed values for days two through six, divided by five. In the same way, you would compute moving averages for all remaining five-day periods through the 200th index value. Note that you would use actual values for the first through fourth days.

The relative percentage data is sure to contain some variability because it is based on performance data from just one set of drafted players. Again, you can use a moving average to eliminate some of the variability. You choose to use a moving average of three draft picks. For example, the moving average value for the third draft pick would be the sum of the relative percentages for the first three draft picks, divided by three. The moving average value for the fourth draft pick would be the sum of the relative percentages for draft picks two through four, divided by three. In the same way, you would compute moving averages through draft pick 100. The moving average value for the first and second draft pick is its relative value. When you finish, the first few rows in the summary worksheet should look like Figure 11-21.

	A	B	C	D	E	F	G
1	Draft Num	In The League	Starter In Year	All Star In Year	Total Points	Relative	MovingAvg
2	1	30	60	90	180	2.77%	2.77%
3	2	30	60	60	150	2.31%	2.31%
4	3	30	60	0	90	1.39%	2.16%
5	4	30	60	0	90	1.39%	1.69%
6	5	30	60	0	90	1.39%	1.39%
7	6	20	40	60	120	1.85%	1.54%

Source: Microsoft product screenshots used with permission from Microsoft Corporation

FIGURE 11-21   Rows in the Summary worksheet after smoothing

Charting the Data

You want to create scatter charts of the coach's data and the actual performance data. Open a new worksheet and rename it Charts. Then copy the data for the draft numbers, coach's relative values, and moving averages to the sheet. The coach's relative values come from the Coaches Values worksheet you created earlier. When you finish copying the data, your worksheet should look like Figure 11-22.

	A	B	C
1	Draft Num	Coaches Relative Value	MovingAvg
2	1	5.28%	2.77%
3	2	4.58%	2.31%
4	3	3.87%	2.16%
5	4	3.17%	1.69%
6	5	2.99%	1.39%
7	6	2.82%	1.54%
8	7	2.64%	1.44%

Source: Microsoft product screenshots used with permission from Microsoft Corporation

FIGURE 11-22   Rows in the Charts worksheet

To chart Coach Smith's data, first select the data in the Draft Num and Coaches Relative Value columns. To create the scatter chart, click the Insert tab, and then click the down arrow on the Scatter charts icon in the Charts group. When you select Scatter Chart with Smooth Lines, the chart will be created automatically by Excel. Your worksheet should look like Figure 11-23.

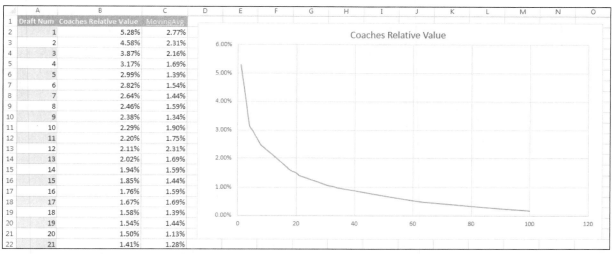

Source: Microsoft product screenshots used with permission from Microsoft Corporation

FIGURE 11-23   Coach's relative value chart

If you click the chart, a plus sign icon (+) appears next to it. Click the + icon to add chart elements. Click the arrow associated with Axis Titles to add appropriate horizontal and vertical titles.

You now want to add the moving average scatter plot in the same chart. Click the chart, and then click the Design tab under Chart Tools. Next, click the Select Data button, and then click the Add button in the Select Data Source window that appears. Enter a series name, such as Moving Average. Click in the box used to enter X values and then select the 100 draft number values, but not the column heading. The box used to enter Y values contains several characters; use the backspace key to delete the characters, but leave the equals sign. Then select the 100 moving average values, but not the column header. Select OK to add the moving average scatter plot. Your worksheet should look like Figure 11-24.

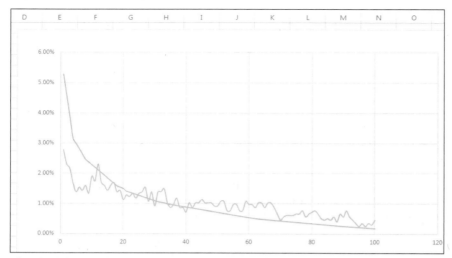

Source: Microsoft product screenshots used with permission from Microsoft Corporation

FIGURE 11-24   Moving average added to chart

Note that Excel might remove the original chart title. You can use the + icon to add another chart title.

Even though you used a moving average, the scatter plot still contains some variability. Fortunately, Excel lets you fit the scatter plot's trend line, which adds further smoothing. Click the moving average line, and then click the + icon. Click the arrow associated with Trendline, select More Options, and then choose a Polynomial trend line. Excel inserts the trend line, as shown in Figure 11-25.

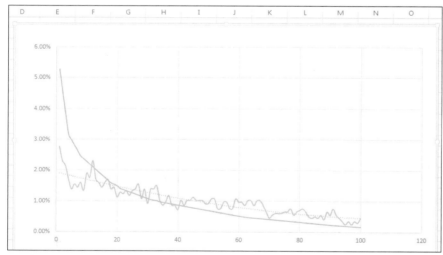

Source: Microsoft product screenshots used with permission from Microsoft Corporation

FIGURE 11-25   Moving average trend line added to chart

The moving average trend line represents the value of actual performance by players drafted three years ago. You can now compare these values with the coach's values. You should be able to see that the coach's assigned values are higher for early draft picks, but the actual values for later picks exceed the coach's values.

Save and then close the Excel file.

ASSIGNMENT 2: DOCUMENTING FINDINGS IN A MEMO

In this assignment, you write a memo in Microsoft Word that documents your findings. The GM wants these questions addressed:

- Do draft picks decrease in actual value as the draft goes on? Refer to your first two queries to answer this question.

- How frequently are good players drafted after the first round? You need not quantify this answer. Develop your answer by scanning the output of your third query.
- Does the coach's chart greatly overvalue early picks? Does the chart greatly undervalue later picks? If there are big differences, how can the GM use this knowledge? Include a copy of your chart in the memo and refer to your scatter plots to answer these questions.

In your memo, observe the following requirements:

- Your memo should have proper headings such as Date, To, From, and Subject. You can address the memo to the team's general manager. Set up the memo as discussed in Tutorial E.
- Briefly outline the situation. However, you need not provide much background—you can assume that readers are generally familiar with your task.
- In the body of the memo, answer the GM's questions.

ASSIGNMENT 3: GIVING AN ORAL PRESENTATION

Your instructor may require that you summarize your analysis and results in an oral presentation. Assume that you would be talking to the GM and her small staff for 10 minutes or less. Use visual aids or handouts as appropriate. See Tutorial F for guidance on preparing and giving an oral presentation.

DELIVERABLES

Assemble the following deliverables for your instructor:

1. Printout of your memo
2. Spreadsheet printouts, if required by your instructor
3. Query printouts, if required by your instructor
4. Electronic media such as flash drive or CD, which should include your Word file, Access file, and Excel file

Staple the printouts together with the memo on top. If you have more than one .xlsx file or .accdb file stored on your electronic media, write your instructor a note that identifies the files for this assignment.

THE FOOTBALL CONCUSSION FREQUENCY ANALYSIS

Decision Support with Microsoft Access and Excel

PREVIEW

Many players from a college football conference in your region have suffered concussions in recent years. Conference officials want to know more about the frequency and incidence of these concussions. In this case, you will use Microsoft Access and Excel to prepare an analysis of concussions that occurred in the past two football seasons.

PREPARATION

- Review database and spreadsheet concepts discussed in class and in your textbook.
- Complete any exercises that your instructor assigns.
- Complete any parts of Tutorials B, C, and D that your instructor assigns, or refer to them as necessary.
- Review file-saving procedures for Windows programs, as discussed in Tutorial C.
- Refer to Tutorials E and F as necessary.

BACKGROUND

Your university has a football team that plays in a conference with seven other college teams in the region. Each team plays a 12-game schedule, seven of them against the other conference teams.

Football is a game that involves frequent violent collisions between players. In these collisions, a player's brain can smash against the inside of his skull, causing an injury called a concussion. In the two prior seasons, concussions suffered by players in your school's conference have been very common. Conference officials know they must do something to address this problem. The first step is to better understand the problem, and you have been called in to help.

Until recent years, football players, coaches, and team doctors thought that concussions were not a serious issue. A player who had been knocked out during a game was expected to come to his senses, gather his thoughts, and then return to the game. However, medical researchers know now that concussions are a very serious medical issue. Symptoms of a concussion, such as disorientation, nausea, severe headaches, poor eyesight, and a general feeling of weakness, can persist indefinitely. Today, a player who suffers a concussion cannot play football again until the symptoms completely abate. Furthermore, a person who suffers a concussion is more likely to suffer another one in the future. Many medical researchers believe that concussions lead to dementia or other mental illness later in life.

Officials in your university's conference are aware of the urgent need to prevent concussions as much as possible. These officials have gathered data about player concussions on the eight conference teams during the last two years. There are many questions about the data, including the following:

- Are players at certain positions more likely than others to suffer a concussion?
- Some coaches have their players practice harder than other coaches, and officials want to know if some practice regimens lead more readily to concussions.
- Football fields are often grass and dirt, and others are made of a harder composite material that is commonly called turf. Officials want to know if certain types of fields correlate with more concussions.

- Is there any evidence that some injuries diagnosed as concussions were actually less severe head injuries?

The data for this case is in the Concussions.accdb database, which is available for you to use; it will save you time. If you want to use this database, locate Case 12 in your data files and then select **Concussions.accdb**.

The tables in the file are discussed next. Figure 12-1 shows the design of the Player table.

Player	
Field Name	Data Type
Player Num	Number
Last Name	Short Text
First Name	Short Text
Position	Short Text
Team	Short Text
Side	Short Text

Source: Microsoft product screenshots used with permission from Microsoft Corporation

FIGURE 12-1   Player table design

The table contains data about conference players who have suffered concussions in the prior two seasons. Each player is assigned a number in the database. The Player Num field is the table's primary key field. The Team and Position fields record the player's team and position played, respectively. The Side field records whether the player is on offense or defense.

Figure 12-2 shows a few of the records in the Player table.

Player Num	Last Name	First Name	Position	Team	Side
1112	Dowd	Reggie	QB	Cruisers	OFF
1113	Ureno	Milford	RB	Destroyers	OFF
1114	Leung	Omar	WR	Roses	OFF
1115	Mclachlan	Clifford	C	Bisons	OFF
1116	Shick	Dwight	OT	Snakes	OFF
1117	Mcbay	Jonah	OG	Vipers	OFF
1118	Repp	Lanny	WR	Geckos	OFF

Source: Microsoft product screenshots used with permission from Microsoft Corporation

FIGURE 12-2   Some data records in the Player table

For example, player 1112 is named Reggie Dowd, a quarterback who plays for the Cruisers. A quarterback is an offensive player.

Data for the eight teams in the conference is recorded in the Teams table. Figure 12-3 shows the design of this table.

Teams	
Field Name	Data Type
Team	Short Text
Home Field	Short Text
Practice	Short Text

Source: Microsoft product screenshots used with permission from Microsoft Corporation

FIGURE 12-3   Design of Teams table

The teams have unique names, so the Team field is the primary key. The Home Field field indicates whether the team's home field has a grass surface or a harder turf surface. The Practice data field indicates how hard the team's coach works the players in practice. A value of Low means the coach makes little use of man-on-man contact drills in practice. A value of High means the coach uses extensive man-on-man contact drills in practice. Figure 12-4 shows the eight records in the Teams table.

Teams

Team	Home Field	Practice
Badgers	Grass	Medium
Bisons	Grass	Medium
Cruisers	Grass	Low
Destroyers	Grass	Low
Geckos	Grass	Medium
Roses	Grass	Medium
Snakes	Turf	High
Vipers	Turf	High

Source: Microsoft product screenshots used with permission from Microsoft Corporation

FIGURE 12-4 Data records in the Teams table

For example, the Geckos play their home games on a grass field. Their coach uses man-on-man contact drills in practice. Some coaches in the conference use more of these drills, and others use less.

The Incidents table records data about each concussion in the past two years. Figure 12-5 shows the design of the table.

Incidents

Field Name	Data Type
Injury Num	Number
Player Num	Number
Games Missed	Number
Game Field	Short Text
Year	Number

Source: Microsoft product screenshots used with permission from Microsoft Corporation

FIGURE 12-5 Design of the Incidents table

Each injury is assigned a unique number. Injury Num is the primary key field. The Player Num field identifies the player who was injured. The Games Missed field records how many games the player missed after the concussion. The Game Field field records the type of playing field (grass or turf) where the injury occurred. The Year field indicates whether the injury occurred this season or in the prior season; a value of 1 represents the prior season and a value of 2 represents the current season.

Figure 12-6 shows a few records in the Incidents table.

Incidents

Injury Num	Player Num	Games Missi	Game Field	Year
1	1112	1	Turf	1
2	1113	1	Turf	1
3	1114	1	Turf	1
4	1115	1	Turf	1
5	1116	1	Turf	1
6	1117	1	Grass	1
7	1118	1	Turf	1
8	1119	1	Turf	1
9	1120	1	Grass	1

Source: Microsoft product screenshots used with permission from Microsoft Corporation

FIGURE 12-6 Data records in the Incidents table

For example, Player 1118 suffered a concussion in a game played on a turf field. He missed one game as a result of the injury, which happened in the prior season. Note that a player might suffer a concussion in the prior season and then another in the current season. Each injury is assigned a number in the database.

ASSIGNMENT 1: USING ACCESS AND EXCEL FOR DECISION SUPPORT

Assume that you are working with the conference's chief medical officer (CMO) to develop your analysis of concussions. The CMO has several questions, which you will attempt to answer using the output of Access queries you design. You will attempt to answer other questions by bringing Access data into Excel and then using pivot tables for further analysis.

Access Database Queries

You will develop six queries in Access.

Query 1

The CMO wants to know if some teams have more concussions than others. Create a query to address that concern. The first part of the query output should look like the data shown in Figure 12-7.

Team	Number Of Injuries	Number of Games Missed
Snakes	56	80
Roses	40	54
Geckos	39	50
Vipers	38	54

FreqByTeam

Source: Microsoft product screenshots used with permission from Microsoft Corporation

FIGURE 12-7 Frequency of injury by team

Sort the output by number of injuries in descending order by team. Save the query as FreqByTeam.

Query 2

The CMO wants to know if players at certain positions have more concussions than others. Create a query to address that question. The first part of the query output should look like the data shown in Figure 12-8.

Position	Number of Injuries
WR	35
OT	29
DT	29
DE	29
LB	27
RB	25

FreqByPosition

Source: Microsoft product screenshots used with permission from Microsoft Corporation

FIGURE 12-8 Frequency of injury by position

The output should be sorted by number of injuries in descending order by position. Save the query as FreqByPosition.

Query 3

The CMO wants some detailed information about players who missed more than two games after a concussion. Create a query that reports this information. The first part of the query output should look like the data shown in Figure 12-9.

SignificantGamesMissed

Player Num	Last Name	Year	Games Missed	Team
1397	Absher	2	5	Snakes
1353	Beals	2	4	Cruisers
1294	Brookshire	2	4	Snakes
1251	Cambron	1	4	Badgers
1251	Cambron	2	3	Badgers
1162	Comacho	1	3	Bisons
1213	Dapolito	1	4	Geckos
1352	Dauber	2	3	Badgers
1391	Emmert	2	4	Geckos
1375	Fehrenbach	2	4	Destroyers
1132	Gillenwater	1	3	Snakes
1244	Groves	2	3	Bisons
1193	Henkel	1	3	Vipers

Source: Microsoft product screenshots used with permission from Microsoft Corporation

FIGURE 12-9 Significant number of games missed

The CMO can scroll the query output to see how frequently players missed more than two games. Note that player 1251 suffered a concussion in each year. Save the query as SignificantGamesMissed.

Query 4

The CMO wants to know how many concussions occurred during each of the past two seasons. Create a query whose output reports the data. The results should look like the data shown in Figure 12-10.

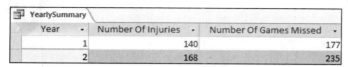

YearlySummary		
Year	Number Of Injuries	Number Of Games Missed
1	140	177
2	168	235

Source: Microsoft product screenshots used with permission from Microsoft Corporation

FIGURE 12-10   Number of injuries each year

Note that more concussions occurred during the season just completed than during the prior season. Save the query as YearlySummary.

Query 5

The CMO wants to know whether offensive or defensive players suffer more concussions. Create a query whose output reports the data. The results should look like the data shown in Figure 12-11.

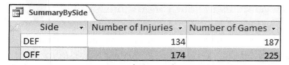

SummaryBySide		
Side	Number of Injuries	Number of Games
DEF	134	187
OFF	174	225

Source: Microsoft product screenshots used with permission from Microsoft Corporation

FIGURE 12-11   Number of offensive and defensive injuries

Save the query as SummaryBySide.

Query 6

You need to create a query that gathers data to be used in Excel. The first few records in the results should look like the data shown in Figure 12-12.

SummaryData								
Injury Num	Player Num	Year	Position	Side	Team	Home Field	Practice	Game Field
1	1112	1	QB	OFF	Cruisers	Grass	Low	Turf
2	1113	1	RB	OFF	Destroyers	Grass	Low	Turf
3	1114	1	WR	OFF	Roses	Grass	Medium	Turf
4	1115	1	C	OFF	Bisons	Grass	Medium	Turf
5	1116	1	OT	OFF	Snakes	Turf	High	Turf
6	1117	1	OG	OFF	Vipers	Turf	High	Grass
7	1118	1	WR	OFF	Geckos	Grass	Medium	Turf
8	1119	1	OT	OFF	Badgers	Grass	Medium	Turf
9	1120	1	DE	DEF	Cruisers	Grass	Low	Grass

Source: Microsoft product screenshots used with permission from Microsoft Corporation

FIGURE 12-12   Data to be used in Excel

Save the query as SummaryData.

Other Queries

If you think other queries and output are needed, feel free to develop them. Your instructor might also specify that you develop more queries to supplement this case.

When you finish creating the queries, close the database file and exit Access.

Importing Summary Data Into Excel

Open a new file in Excel and save it as **Concussions.xlsx**.

Import the SummaryData query output into Excel. First, click the Data tab, and then select From Access in the Get External Data group. Specify the filename of the Access database, and then specify the SummaryData query and where to place the data in the worksheet (cell A1 is recommended).

The data is imported into Excel as a data table, which is not the format that you want. In the Tools group, select Convert To Range. Rename the worksheet as Summary. The first few rows of your worksheet should look like Figure 12-13.

Injury Num	Player Num	Year	Position	Side	Team	Home Field	Practice	Game Field
1	1112	1	QB	OFF	Cruisers	Grass	Low	Turf
2	1113	1	RB	OFF	Destroyers	Grass	Low	Turf
3	1114	1	WR	OFF	Roses	Grass	Medium	Turf
4	1115	1	C	OFF	Bisons	Grass	Medium	Turf
5	1116	1	OT	OFF	Snakes	Turf	High	Turf
6	1117	1	OG	OFF	Vipers	Turf	High	Grass
7	1118	1	WR	OFF	Geckos	Grass	Medium	Turf
8	1119	1	OT	OFF	Badgers	Grass	Medium	Turf

Source: Microsoft product screenshots used with permission from Microsoft Corporation

FIGURE 12-13   Rows in the Summary worksheet

Using Pivot Tables to Gather Data

You will create pivot tables from the Summary worksheet to develop data about concussions in the conference during the last two seasons.

If you need help creating pivot tables, you can refer to Tutorial E, but a short review of the procedure is in order here. First, select a cell in the underlying data range. (Cell A1 is a good choice for your pivot tables.) Select the Insert tab and then select PivotTable in the Tables group. The Create PivotTable window appears, and the proper data range should already be shown. A new worksheet is the default destination; leave that setting and click OK. In the new worksheet that appears, the right side displays the PivotTable Fields window.

The use of this window is best explained by an example. If you want to create a pivot table that displays the total number of injuries in each of the past two years, you would select the Year field and drag it down into the Rows cell. Next, you would select the Injury Num field and drag it into the Values cell. Excel assumes that you want to sum a field's value in the Values cell. In this case, you want to count the number of injuries, not add them up. Select the entry in the Values cell, select Value Field Settings from the resulting menu, and then select Count. Your pivot table would look like the one shown in Figure 12-14.

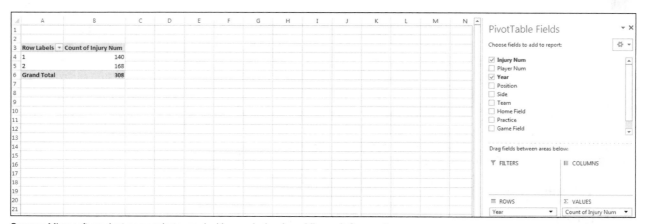

Source: Microsoft product screenshots used with permission from Microsoft Corporation

FIGURE 12-14   Injuries by Year pivot table

Note that the worksheet can be used like any other worksheet. For example, you can use the values in the pivot table cells in formulas elsewhere in the worksheet. Also, you can enter text and format cells, as usual. In Figure 12-15, a header has been entered above the pivot table.

	A	B
1	Injury Count By Year	
2		
3	Row Labels ▼	Count of Injury Num
4	1	140
5	2	168
6	Grand Total	308

Source: Microsoft product screenshots used with permission from Microsoft Corporation

FIGURE 12-15   Injuries by Year pivot table with header

Injuries by Type of Game Field

Create a pivot table that shows injuries by the type of field on which the game was played. Insert an appropriate header above the pivot table. By Excel formula, insert a column that reports the amounts of concussions on each type of field as percentages of total concussion injuries. Your output should look like that shown in Figure 12-16.

	A	B	C
1	Injuries By Type of Game Field		
2			Injury
3	Row Labels ▼	Count of Injury Num	Percent
4	Grass	107	35%
5	Turf	201	65%
6	Grand Total	308	100%

Source: Microsoft product screenshots used with permission from Microsoft Corporation

FIGURE 12-16   Injuries by type of game field

Only 25 percent of conference home fields are turf fields, but many nonconference games are played on turf. Overall, between 40 and 50 percent of the conference's games are played on turf when nonconference games are included in the data.

Add a comment to the header field to explain what the pivot table is intended to show. To add a comment to a cell, right-click it, click Insert Comment, and then type the comment in the text box that appears. (Later, a worksheet user can read the comment by moving the mouse over the cell.)

Injuries by Type of Practice

Create a pivot table that shows injuries by the type of practice regimen coaches use. Insert an appropriate header above the pivot table. By Excel formula, insert a column that displays the amounts of concussions for each type of practice regimen as percentages of total concussion injuries. Enter a column that shows the percentages of each practice regimen used in the conference (25%, 25%, and 50%). Then, by IF statement, insert a column that compares the percentages of concussions for each practice regimen with percentages for how often each type of practice regimen is used throughout the conference. Your output should look like that shown in Figure 12-17.

	A	B	C	D	E
1	Injuries By Type of Practice				
2			Injury	League	Vs. League
3	Row Labels ▼	Count of Injury Num	Percent	Percent	Percent
4	High	94	31%	25%	More
5	Low	64	21%	25%	Less
6	Medium	150	49%	50%	Less
7	Grand Total	308	100%	100%	

Source: Microsoft product screenshots used with permission from Microsoft Corporation

FIGURE 12-17   Injuries by type of practice regimen

Add a comment to the header field to explain what the pivot table is intended to show.

Injuries by Year and by Position

Create a pivot table that reports injuries by the player's position. When setting up the table, drag the Position field into the Rows cell, and drag the Year field into the Columns cell.

Insert an appropriate header above the pivot table. By Excel formula, insert a column that shows the amounts of injuries by position as percentages of total injuries.

Insert a column that shows how many players manned each position at one time during the two-year period. This data is shown in the League Count column in Figure 12-18. For example, each team has two centers, so 16 center positions were manned each season. In two seasons, a total of 32 center positions were available at any one time. By Excel formula, compute the number of players available for each position versus the total positions available. Then, using an IF statement, compare the injury percentages for each position with the percentage of players available at that position. Your results should look like those in Figure 12-18.

As shown in the figure, you can then compare injuries at a position with the total number of players at that position. If you assume that 5 percent of players have a certain position, you would probably not be surprised to learn that 5 percent of all concussions were suffered by players at that position. However, you might be surprised to learn that 10 percent of injuries were suffered by players at that position.

	A	B	C	D	E	F	G	H
1	Injuries By Position and Year							
2								
3	Count of Injury Num	Column Labels ▾			Injury	League	League	Vs. League
4	Row Labels ▾		1	2 Grand Total	Percent	Count	Percent	Percent
5	C		4	4 8	3%	32	4%	Less
6	CB		10	15 25	8%	64	9%	Less
7	DE		12	17 29	9%	64	9%	More
8	DT		12	17 29	9%	64	9%	More
9	FB		5	5 10	3%	32	4%	Less
10	LB		11	16 27	9%	96	13%	Less
11	OG		12	13 25	8%	48	7%	More
12	OT		14	15 29	9%	48	7%	More
13	QB		10	10 20	6%	48	7%	Less
14	RB		12	13 25	8%	48	7%	More
15	SY		10	14 24	8%	64	9%	Less
16	TE		11	11 22	7%	32	4%	More
17	WR		17	18 35	11%	80	11%	More
18	Grand Total		140	168 308	100%	720	100%	
19								
20	Grand Total as Percent of Total Positions in League	43%						

Source: Microsoft product screenshots used with permission from Microsoft Corporation

FIGURE 12-18   Injuries by year and position

Below the pivot table, insert a formula that computes the grand total of injuries as a percentage of all positions available in the past two years. Note that this value does not represent the percentage of players who suffered concussions in the past two years. If a player misses a game, another player is activated from the practice squad. Thus, the number of players during the two-year period is greater than the total of all positions in those two years.

Other Pivot Tables

If you think other pivot tables are needed, feel free to develop them. Your instructor might also specify that you develop more pivot tables to supplement this case.

ASSIGNMENT 2: DOCUMENTING FINDINGS IN A MEMO

In this assignment, you write a memo in Microsoft Word that documents your findings. You need to address the following questions posed by conference officials:

- Are players at some positions significantly more likely than others to suffer a concussion? Are players on offense significantly more susceptible to concussions than players on defense, or vice versa?
- Are players on some teams significantly more likely to suffer concussions than others?
- Do some practice regimens lead much more readily to concussions?
- Do certain types of fields correlate to significantly more concussions?
- Is there any evidence that some injuries diagnosed as concussions were actually less severe head injuries?

The questions can be addressed individually or in combination. Refer to your query output and pivot tables to support your answers.

In your memo, observe the following requirements:

- Your memo should have proper headings such as Date, To, From, and Subject. You can address the memo to conference officials. Set up the memo as discussed in Tutorial E.
- Briefly outline the situation. However, you need not provide much background—you can assume that readers are generally familiar with your task.
- In the body of the memo, answer the questions posed by conference officials.
- Include any graphical support that your instructor requires, such as copies of query output or pivot tables.

ASSIGNMENT 3: GIVING AN ORAL PRESENTATION

Your instructor may require that you summarize your analysis and results in an oral presentation. Assume that you would be talking to conference officials for 10 minutes or less. Use visual aids or handouts as appropriate. See Tutorial F for guidance on preparing and giving an oral presentation.

DELIVERABLES

Assemble the following deliverables for your instructor:

1. Printout of your memo
2. Spreadsheet printouts, if required by your instructor
3. Query printouts, if required by your instructor
4. Electronic media such as flash drive or CD, which should include your Word file, Access file, and Excel file

Staple the printouts together with the memo on top. If you have more than one .xlsx file or .accdb file on your electronic media, write your instructor a note that identifies the files for this assignment.

PART 6

ADVANCED SKILLS
USING EXCEL

TUTORIAL E
Guidance for Excel Cases, 243

TUTORIAL **E**

GUIDANCE FOR EXCEL CASES

The Microsoft Excel cases in this book require the student to write a memorandum that includes a table. Guidelines for preparing a memo in Microsoft Word and instructions for entering a table in a Word document are provided to begin this tutorial. Also, some of the cases in this book require the use of advanced Excel techniques. Those techniques are explained in this tutorial rather than in the cases themselves:

- Using data tables
- Using pivot tables
- Using built-in functions

You can refer to Sheet 1 of Tutorial E_data.xlsx when reading about data tables. Refer to Sheet 2 when reading about pivot tables.

PREPARING A MEMORANDUM IN WORD

A business memo should include proper headings, such as TO, FROM, DATE, and SUBJECT. If you want to use a Word memo template, follow these steps:

1. In Word, click File.
2. Click New.
3. Enter "memos" in the Search for online templates box, and then click the Start searching button.
4. Click a memo template, such as Memo (elegant) or another memo design of your choice, and then click the Create button to start a new memo document.

The first time you do this, you may need to click Download to install the template. You might also have to search for the memo templates.

ENTERING A TABLE INTO A WORD DOCUMENT

Enter a table into a Word document using the following procedure:

1. Click the cursor where you want the table to appear in the Word document.
2. In the Tables group on the Insert tab, click the Table drop-down menu.
3. Click Insert Table.
4. Choose the number of rows and columns.
5. Click OK.

DATA TABLES

An Excel data table is a contiguous range of data that has been designated as a table. Once you make this designation, the table gains certain properties that are useful for data analysis. (Note that in some previous versions of Excel, data tables were called *data lists*.) Suppose you have a list of runners who have completed a race, as shown in Figure E-1.

	A	B	C	D	E	F
1	RUNNER#	LAST	FIRST	AGE	GENDER	TIME (MIN)
2	100	HARRIS	JANE	O	F	70
3	101	HILL	GLENN	Y	M	70
4	102	GARCIA	PEDRO	M	M	85
5	103	HILBERT	DORIS	M	F	90
6	104	DOAKS	SALLY	Y	F	94
7	105	JONES	SUE	Y	F	95
8	106	SMITH	PETE	M	M	100
9	107	DOE	JANE	O	F	100
10	108	BRADY	PETE	O	M	100
11	109	BRADY	JOE	O	M	120
12	110	HEEBER	SALLY	M	F	125
13	111	DOLTZ	HAL	O	M	130
14	112	PEEBLES	AL	Y	M	63

Source: Microsoft product screenshots used with permission from Microsoft Corporation

FIGURE E-1   Data table example

To turn the information into a data table, highlight the data range, including headings, and click the Insert tab. Then click Table in the Tables group. The Create Table window appears, as shown in Figure E-2.

Source: Microsoft product screenshots used with permission from Microsoft Corporation

FIGURE E-2   Create Table window

When you click OK, the data range appears as a table. In the Table Style Options group on the Design tab, click the Total Row check box to add a totals row to the data table. You also can select a light style in the Table Styles list to get rid of the contrasting color in the table's rows. Figure E-3 shows the results.

	A	B	C	D	E	F
1	RUNNER#	LAST	FIRST	AGE	GENDER	TIME (MIN)
2	100	HARRIS	JANE	O	F	70
3	101	HILL	GLENN	Y	M	70
4	102	GARCIA	PEDRO	M	M	85
5	103	HILBERT	DORIS	M	F	90
6	104	DOAKS	SALLY	Y	F	94
7	105	JONES	SUE	Y	F	95
8	106	SMITH	PETE	M	M	100
9	107	DOE	JANE	O	F	100
10	108	BRADY	PETE	O	M	100
11	109	BRADY	JOE	O	M	120
12	110	HEEBER	SALLY	M	F	125
13	111	DOLTZ	HAL	O	M	130
14	112	PEEBLES	AL	Y	M	63
15	Total					1242

Source: Microsoft product screenshots used with permission from Microsoft Corporation

FIGURE E-3   Data table example

The headings have acquired drop-down menu tabs, as you can see in Figure E-3.

You can sort the data table records by any field. Perhaps you want to sort by times. If so, click the drop-down menu in the TIME (MIN) heading, and then click Sort Smallest to Largest. You get the results shown in Figure E-4.

	A	B	C	D	E	F
1	RUNNER ▾	LAST ▾	FIRST ▾	AGE ▾	GENDE ▾	TIME (MIN ▾↑
2	112	PEEBLES	AL	Y	M	63
3	100	HARRIS	JANE	O	F	70
4	101	HILL	GLENN	Y	M	70
5	102	GARCIA	PEDRO	M	M	85
6	103	HILBERT	DORIS	M	F	90
7	104	DOAKS	SALLY	Y	F	94
8	105	JONES	SUE	Y	F	95
9	106	SMITH	PETE	M	M	100
10	107	DOE	JANE	O	F	100
11	108	BRADY	PETE	O	M	100
12	109	BRADY	JOE	O	M	120
13	110	HEEBER	SALLY	M	F	125
14	111	DOLTZ	HAL	O	M	130
15	Total					1242

Source: Microsoft product screenshots used with permission from Microsoft Corporation
FIGURE E-4   Sorting list by drop-down menu

You can see that Peebles had the best time and Doltz had the worst time. You also can sort from Largest to Smallest.

In addition, you can sort by more than one criterion. Assume that you want to sort first by gender and then by time (within gender). You first sort by gender from A to Z. Then you again click the Gender drop-down tab, point to Sort by Color, and then click Custom Sort. In the Sort window that appears, click Add Level and choose Time as the next criterion. See Figure E-5.

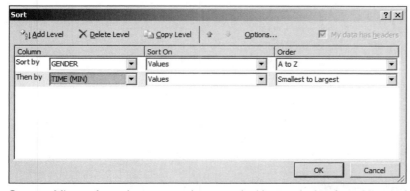

Source: Microsoft product screenshots used with permission from Microsoft Corporation
FIGURE E-5   Sorting on multiple criteria

Click OK to get the results shown in Figure E-6.

	A	B	C	D	E	F
1	RUNNER	LAST	FIRST	AGE	GENDE	TIME (MIN
2	100	HARRIS	JANE	O	F	70
3	103	HILBERT	DORIS	M	F	90
4	104	DOAKS	SALLY	Y	F	94
5	105	JONES	SUE	Y	F	95
6	107	DOE	JANE	O	F	100
7	110	HEEBER	SALLY	M	F	125
8	112	PEEBLES	AL	Y	M	63
9	101	HILL	GLENN	Y	M	70
10	102	GARCIA	PEDRO	M	M	85
11	106	SMITH	PETE	M	M	100
12	108	BRADY	PETE	O	M	100
13	109	BRADY	JOE	O	M	120
14	111	DOLTZ	HAL	O	M	130
15	Total					1242

Source: Microsoft product screenshots used with permission from Microsoft Corporation

FIGURE E-6   Sorting by gender and time (within gender)

You can see that Harris had the best female time and that Peebles had the best male time.

Perhaps you want to see the top *n* listings for some attribute; for example, you may want to see the top five runners' times. Select the Time column's drop-down menu, and select Number Filters. From the menu that appears, click Top 10. The Top 10 AutoFilter window appears, as shown in Figure E-7.

Source: Microsoft product screenshots used with permission from Microsoft Corporation

FIGURE E-7   Top 10 AutoFilter window

This window lets you specify the number of values you want. You might see 10 values as a default setting when the window appears. Figure E-7 shows that the user specified five values. Click OK to get the results shown in Figure E-8.

	A	B	C	D	E	F
1	RUNNER	LAST	FIRST	AGE	GENDE	TIME (MIN
6	107	DOE	JANE	O	F	100
7	110	HEEBER	SALLY	M	F	125
11	106	SMITH	PETE	M	M	100
12	108	BRADY	PETE	O	M	100
13	109	BRADY	JOE	O	M	120
14	111	DOLTZ	HAL	O	M	130
15	Total					675

Source: Microsoft product screenshots used with permission from Microsoft Corporation

FIGURE E-8   Top 5 times

The output contains more than five data records because there are ties at 100 minutes. If you want to see all of the records again, click the Time drop-down menu and click Clear Filter From "TIME (MIN)." The full table of data reappears, as shown in Figure E-9.

	A	B	C	D	E	F
1	RUNNER	LAST	FIRST	AGE	GENDE	TIME (MIN
2	100	HARRIS	JANE	O	F	70
3	103	HILBERT	DORIS	M	F	90
4	104	DOAKS	SALLY	Y	F	94
5	105	JONES	SUE	Y	F	95
6	107	DOE	JANE	O	F	100
7	110	HEEBER	SALLY	M	F	125
8	112	PEEBLES	AL	Y	M	63
9	101	HILL	GLENN	Y	M	70
10	102	GARCIA	PEDRO	M	M	85
11	106	SMITH	PETE	M	M	100
12	108	BRADY	PETE	O	M	100
13	109	BRADY	JOE	O	M	120
14	111	DOLTZ	HAL	O	M	130
15	Total					1242

Source: Microsoft product screenshots used with permission from Microsoft Corporation

FIGURE E-9   Restoring all data to window

Each of the cells in the Total row has a drop-down menu. The menu choices are statistical operations that you can perform on the totals—for example, you can take a sum, take an average, take a minimum or maximum, count the number of records, and so on. Assume that the Time drop-down menu was selected, as shown in Figure E-10. Note that the Sum operator is highlighted by default.

	A	B	C	D	E	F
1	RUNNER	LAST	FIRST	AGE	GENDE	TIME (MIN
2	100	HARRIS	JANE	O	F	70
3	103	HILBERT	DORIS	M	F	90
4	104	DOAKS	SALLY	Y	F	94
5	105	JONES	SUE	Y	F	95
6	107	DOE	JANE	O	F	100
7	110	HEEBER	SALLY	M	F	125
8	112	PEEBLES	AL	Y	M	63
9	101	HILL	GLENN	Y	M	70
10	102	GARCIA	PEDRO	M	M	85
11	106	SMITH	PETE	M	M	100
12	108	BRADY	PETE	O	M	100
13	109	BRADY	JOE	O	M	120
14	111	DOLTZ	HAL	O	M	130
15	Total					1242
16						None
17						Average
						Count
18						Count Numbers
19						Max
						Min
20						Sum
						StdDev
21						Var
22						More Functions...

Source: Microsoft product screenshots used with permission from Microsoft Corporation

FIGURE E-10   Selecting Time drop-down menu in Total row

By changing from Sum to Average, you find that the average time for all runners was 95.5 minutes, as shown in Figure E-11.

	A	B	C	D	E	F
1	RUNNER	LAST	FIRST	AGE	GENDE	TIME (MIN
2	100	HARRIS	JANE	O	F	70
3	103	HILBERT	DORIS	M	F	90
4	104	DOAKS	SALLY	Y	F	94
5	105	JONES	SUE	Y	F	95
6	107	DOE	JANE	O	F	100
7	110	HEEBER	SALLY	M	F	125
8	112	PEEBLES	AL	Y	M	63
9	101	HILL	GLENN	Y	M	70
10	102	GARCIA	PEDRO	M	M	85
11	106	SMITH	PETE	M	M	100
12	108	BRADY	PETE	O	M	100
13	109	BRADY	JOE	O	M	120
14	111	DOLTZ	HAL	O	M	130
15	Total					95.53846154

Source: Microsoft product screenshots used with permission from Microsoft Corporation

FIGURE E-11   Average running time shown in Total row

PIVOT TABLES

Suppose you have data for a company's sales transactions by month, by salesperson, and by amount for each product type. You would like to display each salesperson's total sales by type of product sold and by month. You can use a pivot table in Excel to tabulate that summary data. A pivot table is built around one or more dimensions and thus can summarize large amounts of data. Figure E-12 shows total sales cross-tabulated by salesperson and by month.

	A	B	C	D	E
1	Name	Product	January	February	March
2	Jones	Product1	30,000	35,000	40,000
3	Jones	Product2	33,000	34,000	45,000
4	Jones	Product3	24,000	30,000	42,000
5	Smith	Product1	40,000	38,000	36,000
6	Smith	Product2	41,000	37,000	38,000
7	Smith	Product3	39,000	50,000	33,000
8	Bonds	Product1	25,000	26,000	25,000
9	Bonds	Product2	22,000	25,000	24,000
10	Bonds	Product3	19,000	20,000	19,000
11	Ruth	Product1	44,000	42,000	33,000
12	Ruth	Product2	45,000	40,000	30,000
13	Ruth	Product3	50,000	52,000	35,000

Source: Microsoft product screenshots used with permission from Microsoft Corporation

FIGURE E-12   Excel spreadsheet data

You can create pivot tables and many other kinds of tables with the Excel PivotTable tool. To create a pivot table from the data in Figure E-12, follow these steps:

1.   Starting in the spreadsheet in Figure E-12, click a cell in the data range, and then click the Insert tab. In the Tables group, choose PivotTable. You see the window shown in Figure E-13.

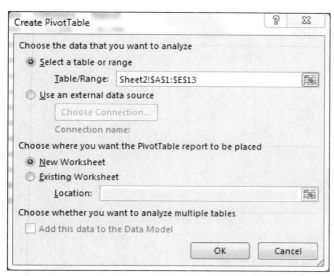

Source: Microsoft product screenshots used with permission from Microsoft Corporation

FIGURE E-13   Creating a pivot table

2.   Make sure New Worksheet is checked under "Choose where you want the PivotTable report to be placed." Click OK. The window shown in Figure E-14 appears.

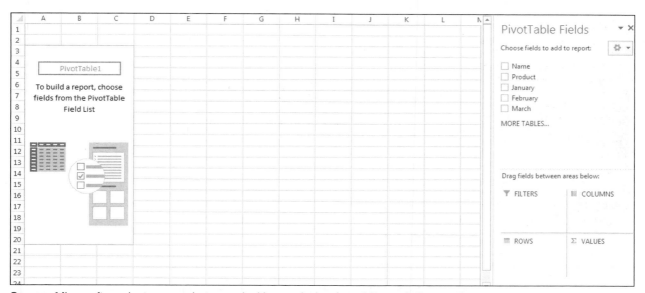

Source: Microsoft product screenshots used with permission from Microsoft Corporation

FIGURE E-14   PivotTable design window

The data range's column headings are shown in the PivotTable Field list on the right side of the window. From there, you can click and drag column headings into the Rows, Columns, and Values panes that appear in the lower-right part of the spreadsheet.

3. If you want to see the January sales by product for each salesperson, drag the Name field to the Columns pane, the Product field to the Rows pane, and the January field to the Values pane. By default, the Sum operation will be shown in the Values pane, which is what you want. You should see the result shown in Figure E-15. Your pivot table should look like the one in Figure E-16.

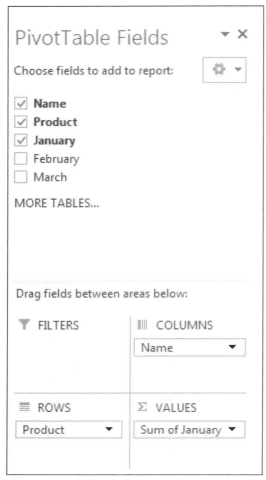

Source: Microsoft product screenshots used with permission from Microsoft Corporation

FIGURE E-15   Pivot table fields

	A	B	C	D	E	F
1						
2						
3	Sum of January	Column Labels				
4	Row Labels	Bonds	Jones	Ruth	Smith	Grand Total
5	Product1	25000	30000	44000	40000	139000
6	Product2	22000	33000	45000	41000	141000
7	Product3	19000	24000	50000	39000	132000
8	Grand Total	66000	87000	139000	120000	412000

Source: Microsoft product screenshots used with permission from Microsoft Corporation

FIGURE E-16   Pivot table

By default, Excel adds all of the sales for each salesperson by month for each product. At the bottom of the pivot table, Excel also shows the total sales for each month for all products.

Note the four small panes in the lower-right corner. The Values pane lets you easily change from the default Sum operator to another one (Min, Max, Average, Count, and so on). Click the drop-down arrow, select Value Field Settings, and then select the desired operator.

BUILT-IN FUNCTIONS

You might need to use some of the following functions when solving the Excel cases elsewhere in this text:

- MIN, MAX, AVERAGE, COUNTIF, ROUND, ROUNDUP, RANDBETWEEN, TREND, and PMT

The syntax of these functions is discussed in this section. The following examples are based on the runner data shown in Figure E-17.

	A	B	C	D	E	F	G
1	RUNNER#	LAST	FIRST	AGE	GENDER	HEIGHT (IN)	TIME (MIN)
2	100	HARRIS	JANE	O	F	60	70
3	101	HILL	GLENN	Y	M	65	70
4	102	GARCIA	PEDRO	M	M	76	85
5	103	HILBERT	DORIS	M	F	64	90
6	104	DOAKS	SALLY	Y	F	62	94
7	105	JONES	SUE	Y	F	64	95
8	106	SMITH	PETE	M	M	73	100
9	107	DOE	JANE	O	F	66	100
10	108	BRADY	PETE	O	M	73	100
11	109	BRADY	JOE	O	M	71	120
12	110	HEEBER	SALLY	M	F	59	125
13	111	DOLTZ	HAL	O	M	76	130
14	112	PEEBLES	AL	Y	M	76	63

Source: Microsoft product screenshots used with permission from Microsoft Corporation

FIGURE E-17   Runner data used to illustrate built-in functions

The data is the same as that shown in Figure E-1, except that Figure E-17 includes a column for the runners' height in inches.

MIN and MAX Functions

The MIN function determines the smallest value in a range of data. The MAX function returns the largest. Say that we want to know the fastest time for all runners, which would be the minimum time in column G. The MIN function computes the smallest value in a set of values. The set of values could be a data range, or it could be a series of cell addresses separated by commas. The syntax of the MIN function is as follows:

- MIN(set of data)

To show the minimum time in cell C16, you would enter the formula shown in the formula bar in Figure E-18:

C16			f_x =MIN(G2:G14)				
	A	B	C	D	E	F	G
1	RUNNER#	LAST	FIRST	AGE	GENDER	HEIGHT	TIME (MIN)
2	100	HARRIS	JANE	O	F	60	70
3	101	HILL	GLENN	Y	M	65	70
4	102	GARCIA	PEDRO	M	M	76	85
5	103	HILBERT	DORIS	M	F	64	90
6	104	DOAKS	SALLY	Y	F	62	94
7	105	JONES	SUE	Y	F	64	95
8	106	SMITH	PETE	M	M	73	100
9	107	DOE	JANE	O	F	66	100
10	108	BRADY	PETE	O	M	73	100
11	109	BRADY	JOE	O	M	71	120
12	110	HEEBER	SALLY	M	F	59	125
13	111	DOLTZ	HAL	O	M	76	130
14	112	PEEBLES	AL	Y	M	76	63
15							
16	MINIMUM TIME:		63				

Source: Microsoft product screenshots used with permission from Microsoft Corporation

FIGURE E-18   MIN function in cell C16

(Assume that you typed the label "MINIMUM TIME:" into cell A16.) You can see that the fastest time is 63 minutes.

To see the slowest time in cell G16, use the MAX function, whose syntax parallels that of the MIN function, except that the largest value in the set is determined. See Figure E-19.

	G16			f_x	=MAX(G2:G14)		
	A	B	C	D	E	F	G
1	RUNNER#	LAST	FIRST	AGE	GENDER	HEIGHT	TIME (MIN)
2	100	HARRIS	JANE	O	F	60	70
3	101	HILL	GLENN	Y	M	65	70
4	102	GARCIA	PEDRO	M	M	76	85
5	103	HILBERT	DORIS	M	F	64	90
6	104	DOAKS	SALLY	Y	F	62	94
7	105	JONES	SUE	Y	F	64	95
8	106	SMITH	PETE	M	M	73	100
9	107	DOE	JANE	O	F	66	100
10	108	BRADY	PETE	O	M	73	100
11	109	BRADY	JOE	O	M	71	120
12	110	HEEBER	SALLY	M	F	59	125
13	111	DOLTZ	HAL	O	M	76	130
14	112	PEEBLES	AL	Y	M	76	63
15							
16	MINIMUM TIME:		63		MAXIMUM TIME:		130

Source: Microsoft product screenshots used with permission from Microsoft Corporation

FIGURE E-19   MAX function in cell G16

AVERAGE, ROUND, and ROUNDUP Functions

The AVERAGE function computes the average of a set of values. Figure E-20 shows the use of the AVERAGE function in cell C17:

	C17			f_x	=AVERAGE(G2:G14)		
	A	B	C	D	E	F	G
1	RUNNER#	LAST	FIRST	AGE	GENDER	HEIGHT	TIME (MIN)
2	100	HARRIS	JANE	O	F	60	70
3	101	HILL	GLENN	Y	M	65	70
4	102	GARCIA	PEDRO	M	M	76	85
5	103	HILBERT	DORIS	M	F	64	90
6	104	DOAKS	SALLY	Y	F	62	94
7	105	JONES	SUE	Y	F	64	95
8	106	SMITH	PETE	M	M	73	100
9	107	DOE	JANE	O	F	66	100
10	108	BRADY	PETE	O	M	73	100
11	109	BRADY	JOE	O	M	71	120
12	110	HEEBER	SALLY	M	F	59	125
13	111	DOLTZ	HAL	O	M	76	130
14	112	PEEBLES	AL	Y	M	76	63
15							
16	MINIMUM TIME:		63		MAXIMUM TIME:		130
17	AVERAGE TIME:		95.53846				

Source: Microsoft product screenshots used with permission from Microsoft Corporation

FIGURE E-20   AVERAGE function in cell C17

Notice that the value shown is a real number with many digits. What if you wanted to have the value rounded to a certain number of digits? Of course, you could format the output cell, but doing that changes only what is shown in the window. You want the cell's contents actually to *be* the rounded number. Therefore, you need to use the ROUND function. Its syntax is:

- ROUND(number, number of digits)

Figure E-21 shows the rounded average time (with two decimal places) in cell G17.

Source: Microsoft product screenshots used with permission from Microsoft Corporation

FIGURE E-21   ROUND function used in cell G17

To achieve this output, cell C17 was used as the value to be rounded. Recall from Figure E-20 that cell C17 had the formula =AVERAGE(G2:G14). The following ROUND formula would produce the same output in cell G17: =ROUND(AVERAGE(G2:G14),2). In this case, Excel evaluates the formula "inside out." First, the AVERAGE function is evaluated, yielding the average with many digits. That value is then input to the ROUND function and rounded to two decimal places.

The ROUNDUP function works much like the ROUND function. ROUNDUP's output is always rounded up to the next value. For example, the value 4 would appear in a cell that contained the following formula: =ROUNDUP(3.12,0). In Figure E-21, if the formula in cell G17 had been =ROUNDUP(AVERAGE(G2:G14),0), the value 96 would have been the result. In other words, 95.54 rounded up with no decimal places becomes 96.

COUNTIF Function

The COUNTIF function counts the number of values in a range that meet a specified condition. The syntax is:

- COUNTIF(range of data, condition)

The condition is a logical expression such as "=1", ">6", or "=F". The condition is shown with quotation marks, even if a number is involved.

Assume that you want to see the number of female runners in cell C18. Figure E-22 shows the formula used.

Source: Microsoft product screenshots used with permission from Microsoft Corporation

FIGURE E-22   COUNTIF function used in cell C18

The logic of the formula is: Count the number of times that "F" appears in the data range E2:E14.

As another example of using COUNTIF, assume that column H shows the rounded ratio of each runner's time in minutes to the runner's height in inches (see Figure E-23).

	H2			f_x	=ROUND(G2/F2,2)			
	A	B	C	D	E	F	G	H
1	RUNNER#	LAST	FIRST	AGE	GENDER	HEIGHT	TIME (MIN)	RATIO
2	100	HARRIS	JANE	O	F	60	70	1.17
3	101	HILL	GLENN	Y	M	65	70	1.08
4	102	GARCIA	PEDRO	M	M	76	85	1.12
5	103	HILBERT	DORIS	M	F	64	90	1.41
6	104	DOAKS	SALLY	Y	F	62	94	1.52
7	105	JONES	SUE	Y	F	64	95	1.48
8	106	SMITH	PETE	M	M	73	100	1.37
9	107	DOE	JANE	O	F	66	100	1.52
10	108	BRADY	PETE	O	M	73	100	1.37
11	109	BRADY	JOE	O	M	71	120	1.69
12	110	HEEBER	SALLY	M	F	59	125	2.12
13	111	DOLTZ	HAL	O	M	76	130	1.71
14	112	PEEBLES	AL	Y	M	76	63	0.83
15								
16	MINIMUM TIME:		63		MAXIMUM TIME:		130	
17	AVERAGE TIME:		95.53846		ROUNDED AVERAGE:		95.54	
18	NUMBER OF FEMALES:		6					

Source: Microsoft product screenshots used with permission from Microsoft Corporation

FIGURE E-23   Ratio of height to time in column H

Assume that all runners whose time in minutes is less than their height in inches will get an award. How many awards are needed? If the ratio is less than 1, an award is warranted. The COUNTIF function in cell G18 computes a count of ratios less than 1, as shown in Figure E-24.

	G18			f_x	=COUNTIF(H2:H14,"<1")			
	A	B	C	D	E	F	G	H
1	RUNNER#	LAST	FIRST	AGE	GENDER	HEIGHT	TIME (MIN)	RATIO
2	100	HARRIS	JANE	O	F	60	70	1.17
3	101	HILL	GLENN	Y	M	65	70	1.08
4	102	GARCIA	PEDRO	M	M	76	85	1.12
5	103	HILBERT	DORIS	M	F	64	90	1.41
6	104	DOAKS	SALLY	Y	F	62	94	1.52
7	105	JONES	SUE	Y	F	64	95	1.48
8	106	SMITH	PETE	M	M	73	100	1.37
9	107	DOE	JANE	O	F	66	100	1.52
10	108	BRADY	PETE	O	M	73	100	1.37
11	109	BRADY	JOE	O	M	71	120	1.69
12	110	HEEBER	SALLY	M	F	59	125	2.12
13	111	DOLTZ	HAL	O	M	76	130	1.71
14	112	PEEBLES	AL	Y	M	76	63	0.83
15								
16	MINIMUM TIME:		63		MAXIMUM TIME:		130	
17	AVERAGE TIME:		95.53846		ROUNDED AVERAGE:		95.54	
18	NUMBER OF FEMALES:		6		RATIOS<1:		1	

Source: Microsoft product screenshots used with permission from Microsoft Corporation

FIGURE E-24   COUNTIF function used in cell G18

RANDBETWEEN Function

If you wanted a cell to contain a randomly generated integer in the range from 1 to 9, you would use the formula =RANDBETWEEN(1,9). Any value between 1 and 9 inclusive would be output by the formula. An example is shown in Figure E-25.

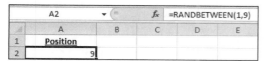

Source: Microsoft product screenshots used with permission from Microsoft Corporation
FIGURE E-25   RANDBETWEEN function used in cell A2

Assume that you copied and pasted the formula to generate a column of 100 numbers between 1 and 9. Every time a value was changed in the spreadsheet, Excel would recalculate the 100 RANDBETWEEN formulas to change the 100 random values. Therefore, you might want to settle on the random values once they are generated. To do this, copy the 100 values, click Paste Special, and then click Values to put the values in the same range. The contents of the cells will change from formulas to literal values.

TREND Function

The TREND function can be used to estimate a variable's value based on the values of other variables. For example, you might know the heights, genders, and weights for 20 people. Correlations exist among these three characteristics. You also have height and gender data for 10 other people, and you want to estimate their weights based on the data you have. The data is shown in Figure E-26.

	A	B	C	D	E	F	G	H	I
1	**Person**	**Height**	**Gender**	**Weight**		**Person**	**Height**	**Gender**	**Pred Weight**
2	101	70	1	190		130	71	1	
3	102	60	2	110		131	61	2	
4	103	72	1	200		132	70	1	
5	104	62	2	120		133	63	2	
6	105	66	1	175		134	65	1	
7	106	66	2	140		135	65	2	
8	107	64	1	170		136	67	1	
9	108	70	2	155		137	70	2	
10	109	62	1	150		138	61	1	
11	110	66	2	150		139	68	2	
12	111	68	1	186					
13	112	68	2	200					
14	113	70	1	200					
15	114	62	2	100					
16	115	72	1	210					
17	116	63	2	110					
18	117	71	1	200					
19	118	64	2	130					
20	119	70	1	170					
21	120	61	2	120					

Source: Microsoft product screenshots used with permission from Microsoft Corporation
FIGURE E-26   Data for people's heights, genders, and weights

The TREND function requires numerical values. In the data, the code for a male is 1 and the code for a female is 2. Height values are measured in inches and weight values are in pounds. For example, person 101 is a male who is 5 feet, 10 inches tall and weighs 190 pounds.

You can use the TREND function to examine a set of data and "learn" the relationship between two or more variables. In this example, the TREND function learns how the heights and genders of 20 people correlate to their weights. Then, given 10 other people's heights and genders, the TREND function applies what it knows to estimate their weights.

The syntax for the TREND function is:

- =TREND(known Ys, known Xs, new Xs)

In the example, the known Ys are the known weights for 20 people, the known Xs are the related heights and genders, and the new Xs are heights and genders of 10 people for whom you want estimated weights. The formula is shown in Figure E-27.

- =TREND($D$2:$D$21, $B$2:$C$21, G2:H2)

Cells D2 to D21 hold the known weights for 20 people. Cells B2 to C21 hold the values of the two predictor variables (height and gender) for those 20 people. Cells G2 and H2 are the predictor variables for person 130, for whom you want a predicted weight. The predicted weight formula is in cell I2.

I2				f_x	=TREND($D$2:$D$21,$B$2:$C$21,G2:H2)				
	A	B	C	D	E	F	G	H	I
1	Person	Height	Gender	Weight		Person	Height	Gender	Pred Weight
2	101	70	1	190		130	71	1	200
3	102	60	2	110		131	61	2	114
4	103	72	1	200		132	70	1	194
5	104	62	2	120		133	63	2	126

Source: Microsoft product screenshots used with permission from Microsoft Corporation

FIGURE E-27   Calculation of predicted weight for person 130

When you copy the formula down the cells in column I for the 10 people, you calculate weight predictions for all of them. By using absolute addressing, the only address changes are the predictor height and gender values for the 10 people.

PMT Function

The PMT function calculates a loan payment. The syntax is:

- =PMT(interest rate, number of periods, initial loan principal)

As an example, assume that you have a 6 percent, 30-year loan for $100,000. The calculation of the monthly payment is shown in Figure E-28.

B5				f_x	=PMT(B1/12,B2*12,B3) * -1		
	A	B	C	D	E		
1	Annual rate:	6%					
2	Years:	30					
3	Principal:	$ 100,000					
4							
5	Monthly Payment:	$599.55					

Source: Microsoft product screenshots used with permission from Microsoft Corporation

FIGURE E-28   Calculation of monthly loan payment

The formula is in cell B5. The monthly interest rate is the annual rate in cell B1 divided by 12. The number of months covered by the loan is the number of years (see cell B2) multiplied by 12. The loan principal is in cell B3. The PMT function returns a negative number, so the expression is multiplied by –1.

Loan payments for the year are computed by multiplying the monthly payment by 12.

PART 7

PRESENTATION SKILLS

TUTORIAL **F**

GIVING AN ORAL PRESENTATION

Giving an oral presentation in class lets you practice the presentation skills you will need in the workplace. The presentations you create for the cases in this textbook will be similar to professional business presentations. You will be expected to present objective, technical results to your organization's stakeholders, and you will have to support your presentation with visual aids commonly used in the business world. During your presentation, your instructor might assign your classmates to role-play an audience of business managers, bankers, or employees. They might also provide feedback on your presentation.

Follow these four steps to create an effective presentation:

1. Plan your presentation.
2. Draft your presentation.
3. Create graphics and other visual aids.
4. Practice delivering your presentation.

PLANNING YOUR PRESENTATION

When planning an oral presentation, you need to know your time limits, establish your purpose, analyze your audience, and gather information. This section explores each of these elements.

Knowing Your Time Limits

You need to consider your time limits on two levels. First, consider how much time you will have to deliver your presentation. For example, what are the key points in your material that can be covered in 10 minutes? The element of time is the primary constraint of any presentation. It limits the breadth and depth of your talk, and the number of visual aids that you can use. Second, consider how much time you will need for the process of preparing your presentation—drafting your presentation, creating graphics, and practicing your delivery.

Establishing Your Purpose

After considering your time limits, you must define your purpose: what you need to say and to whom you will say it. For the Access cases in this book, your purpose will be to inform and explain. For instance, a business's owners, managers, and employees may need to know how the company's database is organized and how they can use it to fill in forms and create reports. In contrast, for the Excel cases, your purpose will be to recommend a course of action based on the results of your business model. You will make the recommendations to business owners, managers, and bankers based on the results of inputting and running various scenarios.

Analyzing Your Audience

Once you have established the purpose of your presentation, you should analyze your audience. Ask yourself: What does my audience already know about the subject? What do the audience members want to know? What do they need to know? Do they have any biases or personal agendas that I should consider? What level of technical detail is best suited to their level of knowledge and interest?

In some Access cases, you will make a presentation to an audience that might not be familiar with Access or with databases in general. In other cases, you might be giving your presentation to a business owner who started to work on a database but was not able to finish it. Tailor the presentation to suit your audience.

For the Excel cases, you are most often interpreting results for an audience of bankers or business managers. In those instances, the audience will not need to know the detailed technical aspects of how you generated your results. But what if your audience consists of engineers or scientists? They will certainly be more interested in the structure and rationale of your decision models. Regardless of the audience, your listeners need to know what assumptions you made prior to developing your spreadsheets because those assumptions might affect their opinion of your results.

Gathering Information

Because you will have just completed a case as you begin preparing your oral presentation, you will already have the basic information you need. For the Access cases, you should review the main points of the case and your goals. Make sure you include all of the points you think are important for the audience to understand. In addition, you might want to go beyond the requirements and explain additional ways in which the database could be used to benefit the organization, now or in the future.

For the Excel cases, you can refer to the tutorials for assistance in interpreting the results from your spreadsheet analysis. For some cases, you might want to use the Internet or the library to research business trends or background information that can support your presentation.

DRAFTING YOUR REPORT AND PRESENTATION

When you have completed the planning stage, you are ready to begin drafting the presentation. At this point, you might be tempted to write your presentation and then memorize it word for word. Even if you could memorize your presentation verbatim, however, your delivery would sound unnatural because people use a simpler vocabulary and shorter sentences when they speak than when they write. For example, read the previous paragraph out loud as if you were presenting it to an audience.

In many business situations, you will be required both to submit a written report of your work and give a PowerPoint presentation. First, write your report, and then design your PowerPoint slides as a "brief" of that report to discuss its main points. When drafting your report and the accompanying PowerPoint slides, follow this sequence:

1. Write the main body of your report.
2. Write the introduction to your report.
3. Write the conclusion to your report.
4. Prepare your presentation (the PowerPoint slides) using your report's main points.

Writing the Main Body

When you draft your report, write the body first. If you try to write the opening paragraph first, you might spend an inordinate amount of time attempting to craft your words perfectly, only to revise the introduction after you write the body of the report.

Keeping Your Audience in Mind

To write the main body, review your purpose and your audience profile. What are the main points you need to make? What are your audience's needs, interests, and technical expertise? It is important to include some technical details in your report and presentation, but keep in mind the technical expertise of your audience.

Remember that the people reading your report or listening to your presentation have their own agendas—put yourself in their places and ask, "What do I need to get out of this presentation?" For example, in the Access cases, an employee might want to know how to enter information on a form, but the business owner might be more interested in generating queries and reports. You need to address their different needs in your presentation. For example, you might say, "And now, let's look at how data entry associates can input data into this form."

Similarly, in the Excel cases, your audience will consist of business owners, managers, bankers, and perhaps some technical professionals. The owners and managers will be concerned with profitability, growth, and customer service. In contrast, the bankers' main concern will be repayment of a loan. Technical professionals will be more concerned with how well your decision model is designed, along with the credibility of the results. You need to address the interests of each group.

Using Transitions and Repetition in your Presentation

During your presentation, remember that the audience is not reading the text of your report, so you need to include transitions to compensate. Words such as *next*, *first*, *second*, and *finally* will help the audience follow the sequence of your ideas. Words such as *however*, *in contrast*, *on the other hand*, and *similarly* will help the audience follow shifts in thought. You can use your voice to convey emphasis.

Also consider using hand gestures to emphasize what you say. For instance, if you list three items, you can use your fingers to tick off each item as you discuss it. Similarly, if you state that profits will be flat, you can make a level motion with your hand for emphasis.

You may be speaking behind a podium or standing beside a projection screen, or both. If you feel uncomfortable standing in one place and you can walk without blocking the audience's view of the screen, feel free to move around. You can emphasize a transition by changing your position. If you tend to fidget, shift, or rock from one foot to the other, try to anchor yourself. A favorite technique of some speakers is to come from behind the podium and place one hand on it while speaking. They get the anchoring effect of the podium while removing the barrier it places between them and the audience. Use the stance or technique that makes you feel most comfortable, as long as your posture or actions do not distract the audience.

As you draft your presentation, repeat key points to emphasize them. For example, suppose your main point is that outsourcing labor will provide the greatest gains in net income. Begin by previewing that concept, and state that you will demonstrate how outsourcing labor will yield the biggest profits. Then provide statistics that support your claim, and show visual aids that graphically illustrate your point. Summarize by repeating your point: "As you can see, outsourcing labor does yield the biggest profits."

Relying on Graphics to Support Your Talk

As you write the main body, think of how to integrate graphics into your presentation. Do not waste words with a long description if a graphic can bring instant comprehension. For instance, instead of describing how information from a query can be turned into a report, show the query and a completed report. Figures F-1 and F-2 illustrate an Access query and the resulting report.

Order Query 1					
Customer Name	City	Product Name	Qty	Price per Unit	Total
Applewood Restaurant	Martinsburg	Frozen Alligator on a Stick	20	$27.99	$559.80
Applewood Restaurant	Martinsburg	Nogales Chipotle Sauce	15	$11.49	$172.35
Applewood Restaurant	Martinsburg	Mom's Deep Dish Apple Pie	12	$12.49	$149.88
Fresh Catch Fishery	Salem	Brumley's Seafood Cocktail Sauce	24	$4.79	$114.96
Fresh Catch Fishery	Salem	NY Smoked Salmon	21	$21.99	$461.79
Fresh Catch Fishery	Salem	Mama Mia's Tiramisu	15	$17.99	$269.85
Jimmy's Crab House	Elkton	Frozen Alligator on a Stick	12	$27.99	$335.88
Jimmy's Crab House	Elkton	Brumley's Seafood Cocktail Sauce	24	$4.79	$114.96
Jimmy's Crab House	Elkton	Mama Mia's Tiramisu	18	$17.99	$323.82
Jimmy's Crab House	Elkton	Mom's Deep Dish Apple Pie	36	$12.49	$449.64

Source: Microsoft product screenshots used with permission from Microsoft Corporation
FIGURE F-1   Access query

Customer Name	City	Product Name	Qty	Price per Unit	Total
Applewood Restaurant	Martinsburg	Frozen Alligator on a Stick	20	$27.99	$559.80
Applewood Restaurant	Martinsburg	Nogales Chipotle Sauce	15	$11.49	$172.35
Applewood Restaurant	Martinsburg	Mom's Deep Dish Apple Pie	12	$12.49	$149.88
Fresh Catch Fishery	Salem	Brumley's Seafood Cocktail Sauce	24	$4.79	$114.96
Fresh Catch Fishery	Salem	NY Smoked Salmon	21	$21.99	$461.79
Fresh Catch Fishery	Salem	Mama Mia's Tiramisu	15	$17.99	$269.85
Jimmy's Crab House	Elkton	Frozen Alligator on a Stick	12	$27.99	$335.88
Jimmy's Crab House	Elkton	Brumley's Seafood Cocktail Sauce	24	$4.79	$114.96
Jimmy's Crab House	Elkton	Mama Mia's Tiramisu	18	$17.99	$323.82
Jimmy's Crab House	Elkton	Mom's Deep Dish Apple Pie	36	$12.49	$449.64

May 2016 Orders--Fine Foods, Inc. — Tuesday, September 02, 2014 11:31:54 AM

Total Orders: $2,952.93

Page 1 of 1

Source: Microsoft product screenshots used with permission from Microsoft Corporation

FIGURE F-2   Access report

Also consider what kinds of graphic media are available and how well you can use them. Your employer will expect you to be able to use Microsoft PowerPoint to prepare your presentation as a slide show. Luckily, many college freshmen are required to take an introductory course that covers Microsoft Office and PowerPoint. If you are not familiar with PowerPoint, several excellent tutorials on the Web can help you learn the basics.

Anticipating the Unexpected

Even though you are only drafting your report and presentation at this stage, eventually you will answer questions from the audience. Being able to handle questions smoothly is the mark of a business professional. The first steps to addressing audience questions are being able to anticipate them and preparing your answers.

You will not use all the facts you gather for your report or presentation. However, as you draft your report, you might want to jot down those facts and keep them handy, in case you need them to answer questions from the audience. PowerPoint has a Notes section where you can include notes for each slide and print them to help you answer questions that arise during your presentation. You will learn how to print notes for your slides later in the tutorial.

The questions you receive depend on the nature of your presentation. For example, during a presentation of an Excel decision model, you might be asked why you are not recommending a certain course of action, or why you left it out of your report. If you have already prepared notes that anticipate such questions, you will probably remember your answers without even having to refer to the notes.

Another potential problem is determining how much technical detail you should display in your slides. In one sense, writing your report will be easier because you can include any graphics, tables, or data you want. Because you have a time limit for your presentation, the question of what to include or leave out becomes more challenging. One approach to this problem is to create more slides than you think you need, and then use the Hide Slide option in PowerPoint to "hide" the extra slides. For example, you might create slides that contain technical details you do not think you will have time to present. However, if you are asked for more details on a particular technical point, you can "unhide" a slide and display the detailed information needed to answer the question. You will learn more about the Hide Slide and Unhide Slide options later in the tutorial.

Writing the Introduction

After you have written the main body of your report and presentation, you can develop the introduction. The introduction should be only a paragraph or two, and it should preview the main points you will cover.

For some of the Access cases, you might want to include general information about databases: what they can do, why they are used, and how they can help a company become more efficient and profitable. You will not need to say much about the business operation because the audience already works for the company.

For the Excel cases, you might want to include an introduction of the general business scenario and describe any assumptions you used to create and run your decision support models. Excel is used for decision support, so you should describe the decision criteria you selected for the model.

Writing the Conclusion

Every good report or presentation needs a good ending. Do not leave the audience hanging. Your conclusion should be brief—only a paragraph or two—and it should give your presentation a sense of closure. Use the conclusion to repeat your main points or, for the Excel cases, to recap your findings and recommendations.

On many occasions, information learned during a business project reveals new opportunities for other projects. Your conclusion should provide closure for the immediate project, but if the project reveals possibilities for future improvements, include them in a "path forward" statement.

CREATING GRAPHICS

Visual aids are a powerful means of getting your point across and making it understandable to your audience. Visual aids come in a variety of forms, some of which are more effective than others. The integrated graphics tools in Microsoft Office can help you prepare a presentation with powerful impact.

Choosing Presentation Media

The media you use will depend on the situation and the media you have available, but remember: *You must maintain control of the media or you will lose the attention of your audience.*

The following list highlights the most common media used in a classroom or business conference room, along with their strengths and weaknesses:

- **PowerPoint slides and a projection system**—These are the predominant presentation media for academic and business use. You can use a portable screen and a simple projector hooked up to a PC, or you can use a full multimedia center. Also, although they are not yet universal in business, touch-sensitive projection screens (for example, Smart Board™ technology) are gaining popularity in college classrooms. The ability to project and display slides, video and sound clips, and live Web pages makes the projection system a powerful presentation tool. *Negatives:* Depending on the complexity of the equipment, you might have difficulties setting it up and getting it to work properly. Also, you often must darken the room to use the projector, and it may be difficult to refer to written notes during your presentation. When using presentation media, you must be able to access and load your PowerPoint file easily. Make sure your file is available from at least two sources that the equipment can access, such as a thumb drive, CD, DVD, or online folder. If your presentation has active links to Web pages, make sure that the presentation computer has Internet access.
- **Handouts**—You can create handouts of your presentation for the audience, which once was the norm for many business meetings. Handouts allow the audience to take notes on applicable slides. If the numbers on a screen are hard to read from the back of the room, your audience can refer to their handouts. With the growing emergence of "green" business practices, however, unnecessary paper use is being discouraged. Many businesses now require reports and presentation slides to be posted at a common site where the audience can access them later. Often, this site is a "public" drive on a business network. *Negatives:* Giving your audience reading material may distract their attention from your presentation. They could read your slides and possibly draw wrong conclusions from them before you have a chance to explain them.

- **Overhead transparencies**—Transparencies are rarely used anymore in business, but some academics prefer them, particularly if they have to write numbers, equations, or formulas on a display large enough for students to see from the back row in a lecture hall. *Negatives:* Transparencies require an overhead projector, and frequently their edges are visually distorted due to the design of the projector lens. You have to use special transparency sheets in a photocopier to create your slides. For both reasons, it is best to avoid using overheads.
- **Whiteboards**—Whiteboards are common in both the business conference room and the classroom. They are useful for posting questions or brainstorming, but you should not use one in your presentation. *Negatives:* You have to face away from your audience to use a whiteboard, and if you are not used to writing on one, it can be difficult to write text that is large enough and legible. Use whiteboards only to jot down questions or ideas that you will check on after the presentation is finished.
- **Flip charts**—Flip charts (also known as easel boards) are large pads of paper on a portable stand. They are used like whiteboards, except that you do not erase your work when the page is full— you flip over to a fresh sheet. Like whiteboards, flip charts are useful for capturing questions or ideas that you want to research after the presentation is finished. Flip charts have the same negatives as whiteboards. Their one advantage is that you can tear off the paper and take it with you when you leave.

Creating Graphs and Charts

Strictly speaking, charts and graphs are not the same thing, although many graphs are referred to as charts. Usually charts show relationships and graphs show change. However, Excel makes no distinction and calls both entities *charts*.

Charts are easy to create in Excel. Unfortunately, the process is so easy that people frequently create graphics that are meaningless, misleading, or inaccurate. This section explains how to select the most appropriate graphics.

You should use pie charts to display data that is related to a whole. For example, you might use a pie chart when breaking down manufacturing costs into Direct Materials, Direct Labor, and Manufacturing Overhead, as shown in Figure F-3. (Note that when you create a pie chart, Excel 2013 will convert the numbers you want to graph into percentages of 100.)

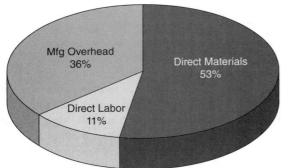

LCD TV Manufacturing Cost

Source: © 2015 Cengage Learning®

FIGURE F-3   3D Pie chart: appropriate use

You would *not*, however, use a pie chart to display a company's sales over a three-year period. For example, the pie chart in Figure F-4 is meaningless because it is not useful to think of the period "as a whole" or the years as its "parts."

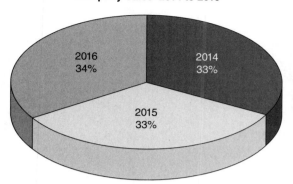

Company Sales–2014 to 2016

Source: © 2015 Cengage Learning®

FIGURE F-4   3D Pie chart: inappropriate use

You should use vertical bar charts (also called column charts) to compare several amounts at the same time, or to compare the same data collected for successive periods of time. The same type of company sales data shown incorrectly in Figure F-4 can be compared correctly using a vertical bar chart (see Figure F-5).

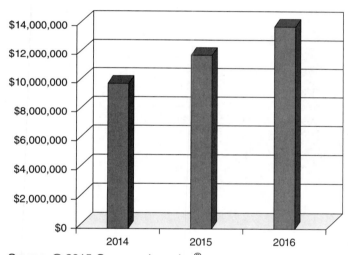

Company Sales–2014 to 2016

Source: © 2015 Cengage Learning®

FIGURE F-5   3D Column chart: appropriate use

As another example, you might want to compare the sales revenues from several different products. You can use a clustered bar chart to show changes in each product's sales over time, as in Figure F-6. This type of bar chart is called a "clustered column" chart in Excel.

When building a chart, include labels that explain the graphics. For instance, when using a graph with an x- and y-axis, you should show what each axis represents so your audience does not puzzle over the graphic while you are speaking. Figures F-6 and F-7 illustrate the necessity of good labels.

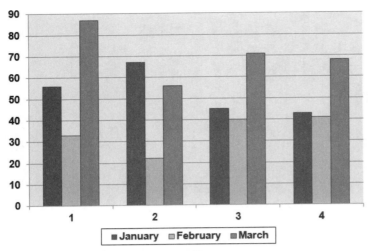

FIGURE F-6   Clustered column graph without title or axis labels

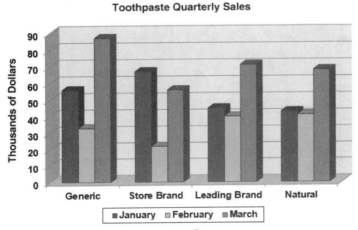

FIGURE F-7   3D Clustered column graph with title and axis labels

In Figure F-6, the graph has no title and neither axis is labeled. Are the amounts in units or dollars? What elements are represented by each cluster of bars? In contrast, Figure F-7 provides a comprehensive snapshot of product sales, which would support a talk rather than create confusion. Note also how the 3D chart style adds visual depth to the chart. Using the 3D chart, the audience can more easily discern that February sales were lower across all product categories.

Another common pitfall of visual aids is charts that have a misleading premise. For example, suppose you want to show how sales are distributed among your inventory, and their contribution to net income. If you simply take the number of items sold in a given month, as displayed in Figure F-8, the visual fails to give your audience a sense of the actual dollar value of those sales. It is far more appropriate and informative to graph the net income for the items sold instead of the number of items sold. The graph in Figure F-9 provides a more accurate picture of which items contribute the most to net income.

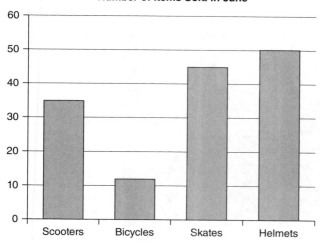

Source: © 2015 Cengage Learning®

FIGURE F-8   Graph of number of items sold that does not reflect generated income

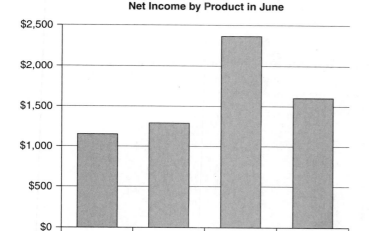

Source: © 2015 Cengage Learning®

FIGURE F-9   Graph of net income by item sold

You should also avoid putting too much data in a single comparative chart. For example, assume that you want to compare monthly mortgage payments for two loan amounts with different interest rates and time frames. You have a spreadsheet that computes the payment data, as shown in Figure F-10.

	A	B	C	D	E	F	G
1	**Calculation of Monthly Payment**						
2	Rate	6.00%	6.10%	6.20%	6.30%	6.40%	6.50%
3	Amount	$ 100,000	$ 100,000	$ 100,000	$ 100,000	$ 100,000	$ 100,000
4	Payment (360 Payments)	$ 599	$ 605	$ 612	$ 618	$ 625	$ 632
5	Payment (180 Payments)	$ 843	$ 849	$ 854	$ 860	$ 865	$ 871
6	Amount	$ 150,000	$ 150,000	$ 150,000	$ 150,000	$ 150,000	$ 150,000
7	Payment (360 Payments)	$ 899	$ 908	$ 918	$ 928	$ 938	$ 948
8	Payment (180 Payments)	$ 1,265	$ 1,273	$ 1,282	$ 1,290	$ 1,298	$ 1,306

Source: Microsoft product screenshots used with permission from Microsoft Corporation

FIGURE F-10   Calculation of monthly payment

In Excel, it is possible (but not advisable) to capture all of the information in a single clustered column chart, as shown in Figure F-11.

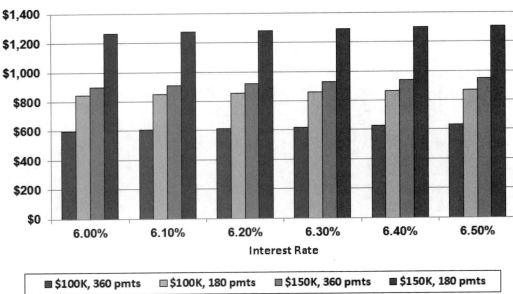

FIGURE F-11   Too much information in one chart

The chart contains a great deal of information. Putting the $100,000 and $150,000 loan payments in the same "cluster" may confuse the readers. They would probably find it easier to understand one chart that summarizes the $100,000 loan (see Figure F-12) and a second chart that covers the $150,000 loan.

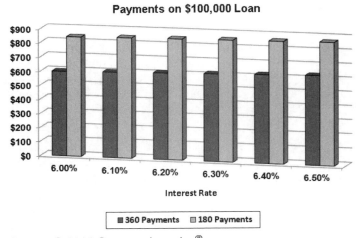

FIGURE F-12   Good balance of information and visual depth

You could then augment the charts with text that summarizes the main differences between the payments for each loan amount. In that fashion, the reader is led step by step through the analysis.

Excel 2007 and later versions no longer have a Chart Wizard; instead, the Insert tab includes a Charts group. Once you create a chart and click it, two chart-specific tabs appear under the Chart Tools heading on the Ribbon to assist you with chart design and formatting. Excel 2013 also adds three menu buttons to the right of the chart: Chart Elements, Chart Styles, and Chart Filters (see Figure F-13). Click each button to see a menu that helps you edit your chart. If you are unfamiliar with the charting tools in Excel, ask your instructor for guidance or refer to the many Excel tutorials on the Web.

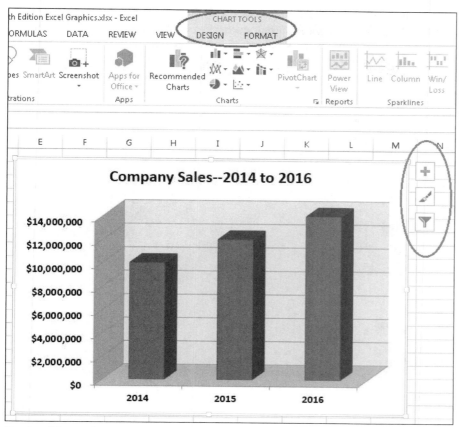

Source: Microsoft product screenshots used with permission from Microsoft Corporation

FIGURE F-13   Chart tools and new menu buttons in Excel 2013

Creating PowerPoint Presentations

PowerPoint presentations are easy to create. When you open PowerPoint, click Blank Presentation. You can select from many different themes, styles, and slide layouts by clicking the Design tab. If none of PowerPoint's default themes suit you, you can download theme "templates" from Microsoft Office Online. When choosing a theme and style for your slides, such as background colors or graphics, fonts, and fills, keep the following guidelines in mind:

- In older versions of PowerPoint, users were advised to avoid pastel backgrounds or theme colors and to keep their slide backgrounds dark. Because of the increasing quality of graphics in both computer hardware and projection systems, most of the default themes in PowerPoint will project well and be easy to read.

- If your projection screen is small or your presentation room is large, consider using boldface type for all of your text to make it readable from the back of the room. If you have time to visit the presentation site beforehand, bring your PowerPoint file, project a slide on the screen, and look at it from the back row of the audience area. If you can read the text, the font is large enough.

- Use transitions and animations to keep your presentation lively, but do not go overboard with them. Swirling letters and pinwheeling words can distract the audience from your presentation.

- It is an excellent idea to animate the text on your slides with entrance effects so that only one bullet point appears at a time when you click the mouse (or when you tap the screen using a touch-sensitive board). This approach prevents your audience from reading ahead of the bullet point being discussed and keeps their attention on you. Entrance effects can be incorporated and managed using the Add Animation button in PowerPoint 2013, as shown in Figures F-14 and F-15.

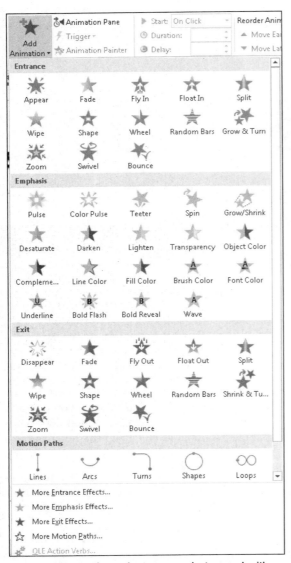

Source: Microsoft product screenshots used with permission from Microsoft Corporation

FIGURE F-14   The Add Animation button on the Ribbon in PowerPoint 2013

FIGURE F-15   Add Entrance Effect window

NOTE—DIFFERENCES IN POWERPOINT ANIMATION TOOLS IN LAST THREE VERSIONS

The structure of the animation tools changed considerably from PowerPoint 2007 to the 2010 version. The Custom Animation button and pane were removed, and most of the custom animation tools were incorporated using the Add Animation button in PowerPoint 2010. You can still use an animation pane to organize and edit your animations within a slide. PowerPoint 2013 has the same Animations tab and Advanced Animations group as PowerPoint 2010.

- Consider creating PowerPoint slides that have a section for your notes. You can print the notes from the Print dialog box by choosing Notes Pages from the Print menu, as shown in Figure F-16. Each slide will be printed at half its normal size, and your notes will appear beneath each slide, as shown by the print preview on the right side of Figure F-16.

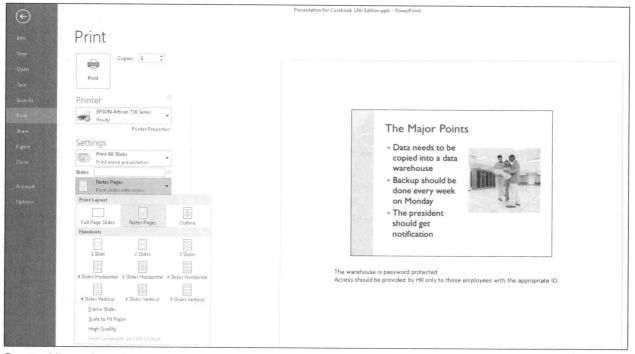

FIGURE F-16   Printing notes page and slide print preview in PowerPoint 2013

- Finally, you should check your PowerPoint slides on a projection screen before your presentation. Information that looks good on a computer display may not be readable on the projection screen.

Using Visual Aids Effectively

Make sure you choose the visual aids that will work most effectively, and that you have enough without using too many. How many is too many? The amount of time you have to speak will determine the number of visual aids you should use, as will your target audience. A good rule of thumb is to allow at least one minute to present each PowerPoint slide. Leave a minimum of two minutes for audience questions after a 10-minute presentation, and allow up to 25 percent of your total presentation time to address questions after longer presentations. (For example, for a 20-minute presentation, figure on taking five minutes for questions.) For a 10-minute talk, try to keep the body of your presentation to eight slides or less. Your target audience will also influence your selection of visual aids. For instance, your slides will need more graphics and animation if you are addressing a group of teenagers than if you are presenting to a board of directors. Remember to use visual aids to emphasize your main points, not to detract from them.

Review each of your slides and visual aids to make sure it meets the following criteria:

- The font size of the text is large enough to read from the back of the presentation area.
- The slide or visual aid does not contain misleading graphics, typographical errors, or misspelled words—the quality of your work is a direct reflection on you.
- The content of your visual aid is relevant to the key points of your presentation.
- The slide or visual aid does not detract from your message. Your animations, pictures, and sound effects should support the text. Your visuals should look professional.
- A visual aid should look good in the presentation environment. If possible, rehearse your PowerPoint slides beforehand in the room where you will give the presentation. Make sure you can read your slides easily from the back row of seats in the room. If you have a friend who can sit in, ask her or him to listen to your voice from the back row of seats. If you have trouble projecting your voice clearly, consider using a microphone for your presentation.
- All numbers should be rounded unless decimals or pennies are crucial. For example, your company might only pay fractions of a cent per Web hit, but this cost may become significant after millions of Web hits.
- Slides should not look too busy or crowded. Many PowerPoint experts have a "6 by 6" rule for bullet points on a slide, which means you should include no more than six bullet points per slide and no more than six words per bullet point. Also avoid putting too many labels or pictures on a slide. Clip art can be "cutesy" and therefore has no place in a professional business presentation. A well-selected picture or two can add emphasis to the theme of a slide. For examples of a slide that is too busy versus one that conveys its points succinctly, see Figures F-17 and F-18.

Major Points

- Data needs to be copied into a data warehouse
- Backup should be done every week on Monday
- The president should get notification
- The vice president should get notification
- The data should be available on the Web
- Web access should be on a secure server
- HR sets passwords
- Only certain personnel in HR can set passwords
- Users need to show ID to obtain a password
- ID cards need to be the latest version

Source: Microsoft product screenshots used with permission from Microsoft Corporation
FIGURE F-17   Busy slide

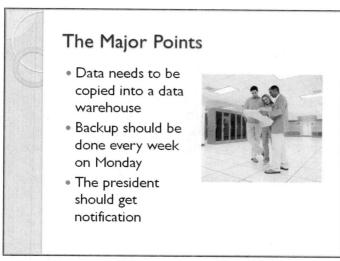

Source: Microsoft product screenshots used with permission from Microsoft Corporation
FIGURE F-18   Slide with appropriate number of bullet points and a supporting photo

You may find that you have created more slides than you have time to present, and you are unsure of which slides you should delete. Some may have data that an audience member might ask about. Fortunately, PowerPoint lets you "hide" slides; these hidden slides will not be displayed in Slide Show view unless you "unhide" them in Normal view. Hiding slides is an excellent way to keep detailed data handy in case your audience asks to see it. Figure F-19 shows how to hide a slide in a PowerPoint presentation. Right-click the slide you want to hide, and then click Hide Slide from the menu to mark the slide as hidden in the presentation. To unhide the slide, right-click it and then click Hide Slide again from the menu. Click the slide to display it in Slide Show view.

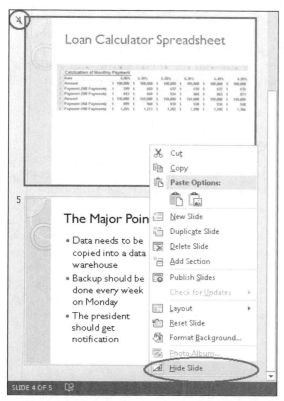

Source: Microsoft product screenshots used with permission from Microsoft Corporation
FIGURE F-19   Hiding a slide in PowerPoint

PRACTICING YOUR DELIVERY

Surveys indicate that public speaking is the greatest fear of many people. However, fear or nervousness can be channeled into positive energy to do a good job. Remember that an audience is not likely to think you are nervous unless you fidget or your voice cracks. Audience members want to hear what you have to say, so think about them and their interests—not about how you feel.

Your presentations for the cases in this textbook will occur in a classroom setting with 20 to 40 students. Ask yourself: Am I afraid when I talk to just one or two of my classmates? The answer is probably no. In addition, they will all have to give presentations as well. Think of your presentation as an extended conversation with several classmates. Let your gaze move from person to person, making brief eye contact with each of them randomly as you speak. As your focus moves from one person to another, think to yourself: I am speaking to one person at a time. As you become more proficient in speaking before a group, your gaze will move naturally among audience members.

Tips for Practicing Your Delivery

Giving an effective presentation is not the same as reading a report to an audience. You should rehearse your message well enough so that you can present it naturally and confidently, with your slides or other visual aids smoothly intermingled with your speaking. The following tips will help you hone the effectiveness of your delivery:

- Practice your presentation several times, and use your visual aids when you practice.
- Show your slides at the right time. Luckily, PowerPoint makes this easy; you can click the slide when you are ready to talk about it. Use cues as necessary in your speaker's notes.
- Maintain eye and voice contact with the audience when using the visual aid. Do not turn your back on your audience. It is acceptable to turn sideways to glance at your slide. A popular trick of experienced speakers is to walk around and steal a glance at the slide while moving to a new position.
- Refer to your visual aids in your talk and use hand gestures where appropriate. Do not ignore your own visual aid, but do not read it to your audience—they can read for themselves.
- Keep in mind that your slides or visual aids should support your presentation, not *be* the presentation. Do not try to crowd the slide with everything you plan to say. Use the slides to illustrate key points and statistics, and fill in the rest of the content with your talk.
- Check your time, especially when practicing. If you stay within the time limit when practicing, you will probably finish a minute or two early when you actually give the presentation. You will be a little nervous and will talk a little faster to a live audience.
- Use numbers effectively. When speaking, use rounded numbers; otherwise, you will sound like a computer. Also make numbers as meaningful as possible. For example, instead of saying "in 83 percent of cases," say "in five out of six cases."
- Do not extrapolate, speculate, or otherwise "reach" to interpret the output of statistical models. For example, suppose your Excel model has many input variables. You might be able to point out a trend, but often you cannot say with mathematical certainty that if a company employs the inputs in the same combination, it will get the same results.
- Some people prefer recording their presentation and playing it back to evaluate themselves. It is amazing how many people are shocked when they hear their recorded voice—and usually they are not pleased with it. In addition, you will hear every *um, uh, well, you know,* throat-clearing noise, and other verbal distraction in your speech. If you want feedback on your presentation, have a friend listen to it.
- If you use a pointer, be careful where you wave it. It is not a light saber, and you are not Luke Skywalker. Unless you absolutely have to use one to point at crucial data on a slide, leave the pointer home.

Handling Questions

Fielding questions from an audience can be tricky because you cannot anticipate all of the questions you might be asked. When answering questions from an audience, *treat everyone with courtesy and respect.* Use the following strategies to handle questions:

- Try to anticipate as many questions as possible, and prepare answers in advance. Remember that you can gather much of the information to prepare those answers while drafting your presentation. The Notes section under each slide in PowerPoint is a good place to write anticipated questions and your answers. Hidden slides can also contain the data you need to answer questions about important details.

- Mention at the beginning of your talk that you will take questions at the end of the presentation, which helps prevent questions from interrupting the flow and timing of your talk. In fact, many PowerPoint presentations end with a Questions slide. If someone tries to interrupt, say that you will be happy to answer the question when you are finished, or that the next graphic answers the question. Of course, this point does not apply to the company CEO—you *always* stop to answer the CEO's questions.

- When answering a question, a good practice is to repeat the question if you have any doubt that the entire audience heard it. Then deliver your answer to the whole audience, but make sure you close by looking directly at the person who asked the question.

- Strive to be informative, not persuasive. In other words, use facts to answer questions. For instance, if someone asks your opinion about a given outcome, you might show an Excel slide that displays the Solver's output; then you can use the data as the basis for answering the question. In that light, it is probably a good idea to have computer access to your Excel model or Access database if your presentation venue permits it, but avoid using either unless you absolutely need it.

- If you do not know the answer to a question, it is acceptable to say so, and it is certainly better than trying to fake the answer. For instance, if someone asks you the difference between the Simplex LP and GRG solving methods in Excel Solver, you might say, "That is an excellent question, but I really don't know the answer—let me research it and get back to you." Then follow up after the presentation by researching the answer and contacting the person who asked the question.

- Signal when you are finished. You might say that you have time for one more question. Wrap up the talk yourself and thank your audience for their attention.

Handling a "Problem" Audience

A "problem" audience or a heckler is every speaker's nightmare. Fortunately, this experience is rare in the classroom: Your audience will consist of classmates who also have to give presentations, and your instructor will be present to intervene in case of problems.

Heckling can be a common occurrence in the political arena, but it does not happen often in the business world. Most senior managers will not tolerate unprofessional conduct in a business meeting. However, fellow business associates might challenge you in what you perceive as a hostile manner. If so, remain calm, be professional, and rely on facts. The rest of the audience will watch to see how you react—if you behave professionally, you make the heckler appear unprofessional by comparison and you'll gain the empathy of the audience.

A more common problem is a question from an audience member who lacks technical expertise. For instance, suppose you explained how to enter data into an Access form, but someone did not understand your explanation. Ask the questioner what part of the explanation was confusing. If you can answer the question briefly and clearly, do so. If your answer turns into a time-consuming dialogue, offer to give the person a one-on-one explanation after the presentation.

Another common problem is receiving a question that you have already answered. The best solution is to give the answer again, as briefly as possible, using different words in case your original answer confused the person. If someone persists in asking questions that have obvious answers, you might ask the audience, "Who would like to answer that question?" The questioner should get the hint.

PRESENTATION TOOLKIT

You can use the form in Figure F-20 for preparation, the form in Figure F-21 for evaluation of Access presentations, and the form in Figure F-22 for evaluation of Excel presentations.

Preparation Checklist

Facilities and Equipment

☐ The room contains the equipment that I need.
☐ The equipment works and I've tested it with my visual aids.
☐ Outlets and electrical cords are available and sufficient.
☐ All the chairs are aligned so that everyone can see me and hear me.
☐ Everyone will be able to see my visual aids.
☐ The lights can be dimmed when/if needed.
☐ Sufficient light will be available so I can read my notes when the lights are dimmed.

Presentation Materials

☐ My notes are available, and I can read them while standing up.
☐ My visual aids are assembled in the order that I'll use them.
☐ A laser pointer or a wand will be available if needed.

Self

☐ I've practiced my delivery.
☐ I am comfortable with my presentation and visual aids.
☐ I am prepared to answer questions.
☐ I can dress appropriately for the situation.

Source: © 2015 Cengage Learning®
FIGURE F-20   Preparation checklist

Evaluating Access Presentations

Course: _____   Speaker: _____   Date: _____

Rate the presentation by these criteria:
4=Outstanding   3=Good   2=Adequate   1=Needs Improvement
N/A=Not Applicable

Content

_____   The presentation contained a brief and effective introduction.

_____   Main ideas were easy to follow and understand.

_____   Explanation of database design was clear and logical.

_____   Explanation of using the form was easy to understand.

_____   Explanation of running the queries and their output was clear.

_____   Explanation of the report was clear, logical, and useful.

_____   Additional recommendations for database use were helpful.

_____   Visuals were appropriate for the audience and the task.

_____   Visuals were understandable, visible, and correct.

_____   The conclusion was satisfying and gave a sense of closure.

Delivery

_____   Was poised, confident, and in control of the audience

_____   Made eye contact

_____   Spoke clearly, distinctly, and naturally

_____   Avoided using slang and poor grammar

_____   Avoided distracting mannerisms

_____   Employed natural gestures

_____   Used visual aids with ease

_____   Was courteous and professional when answering questions

_____   Did not exceed time limit

Submitted by: _____

Source: © 2015 Cengage Learning®

FIGURE F-21   Form for evaluation of Access presentations

Evaluating Excel Presentations

Course: _____ **Speaker:** _____   **Date:** _____

Rate the presentation by these criteria:
4=Outstanding   3=Good   2=Adequate   1=Needs Improvement
N/A=Not Applicable

Content

_____ The presentation contained a brief and effective introduction.

_____ The explanation of assumptions and goals was clear and logical.

_____ The explanation of software output was logically organized.

_____ The explanation of software output was thorough.

_____ Effective transitions linked main ideas.

_____ Solid facts supported final recommendations.

_____ Visuals were appropriate for the audience and the task.

_____ Visuals were understandable, visible, and correct.

_____ The conclusion was satisfying and gave a sense of closure.

Delivery

_____ Was poised, confident, and in control of the audience

_____ Made eye contact

_____ Spoke clearly, distinctly, and naturally

_____ Avoided using slang and poor grammar

_____ Avoided distracting mannerisms

_____ Employed natural gestures

_____ Used visual aids with ease

_____ Was courteous and professional when answering questions

_____ Did not exceed time limit

Submitted by: _____

Source: © 2015 Cengage Learning®

FIGURE F-22 Form for evaluation of Excel presentations